SAP Excellence

Series editors:

Prof. Dr. Dr. h.c. Peter Mertens
Universität Erlangen-Nürnberg

Dr. Peter Zencke,
SAP AG, Walldorf

Springer
Berlin
Heidelberg
New York
Barcelona
Hong Kong
London
Milan
Paris
Singapore
Tokyo

Peter Buxmann
Wolfgang König

Inter-organizational Cooperation with SAP Systems

Perspectives on Logistics
and Service Management

With 92 Figures
and 1 Table

Springer

Prof. Dr. Peter Buxmann
Technical University of Freiberg
Chair of Information Management
Lessingstrasse 45
09599 Freiberg
Germany
buxmann@bwl.tu-freiberg.de

Prof. Dr. Wolfgang König
University of Frankfurt
Institute of Information Systems
Mertonstrasse 17
D-60054 Frankfurt
Germany
wkoenig@wiwi.uni-frankfurt.de

ISBN 3-540-66983-3 Springer-Verlag Berlin Heidelberg New York

Library of Congress Cataloging-in-Publication Data
Die Deutsche Bibliothek – CIP-Einheitsaufnahme
Buxmann, Peter: Inter-organizational Cooperation with SAP Systems: Perspectives on
Logistics and Service Management; with 1 table / Peter Buxmann; Wolfgang König. –
Berlin; Heidelberg; New York; Barcelona; Hong Kong; London; Milan; Paris; Singapore;
Tokyo: Springer, 2000
  (SAP Excellence)
  ISBN 3-540-66983-3

Springer-Verlag Berlin Heidelberg New York
a member of BertelsmannSpringer Science+Business Media GmbH

© Springer-Verlag Berlin · Heidelberg 2000
Printed in Germany

Hardcover-Design: Erich Kirchner, Heidelberg

SPIN 10756988   42/2202-5 4 3 2 1 0 – Printed on acid-free paper

Preface

The formation of *inter-organizational cooperation* is increasingly used to improve the partners´ competitive position in a global world economy. This tendency in logistics occurs when members of a supply chain become partners: suppliers, manufacturers, logistics service providers, and finally the end-customers. *Supply chain management* attempts to optimize the flows of goods and information between companies.

Logistics service companies play an increasingly important role in the supply chain. Their task here is no longer restricted to just providing basic logistics services such as transport, warehousing and transshipment. Rather, logistics service providers are increasingly becoming complete providers of *service and information*. For example, they provide basic services, such as financing for inventory or offer after-sales services. Because of the fundamental importance of the exchange of information for the coordination of the supply chain, logistics service companies are also increasingly becoming information and communications systems specialists.

To support logistics processes, *SAP* provides various systems for the parties in the supply chain. These include modules for the logistics functions in the R/3 System, such as sales and distribution (SD module) and materials management (MM module), and EDI and Workflow Management to link companies using the Internet. In addition, SAP provides various components to optimize the supply chain.

This book concentrates on the business processes linking companies and investigates the opportunities and limits on the use of SAP systems. This book compares the known task requirements to the SAP methodology. Special attention is paid to how SAP supports the function of logistics service providers.

Although this book aims to provide a practical presentation of these concepts and solutions, it does not ignore the scientific foundation. Whereas Chapters 1 and 2 concentrate on providing a compact representation of the available method knowledge, Chapters 3 to 6 show solutions based on SAP systems. The practical orientation is enhanced by the inclusion of case studies: The Schenker Logistics case shows how logistics service providers are increasingly changing to become complete service providers and specialists in the provision and use of information and communications systems (Chapter 7). We use the example of Goodyear to show how modern information and communications systems can support the coordination of logistics between companies (Chapter 8). The use of SAP R/3 and the Supply Chain Management Initiative are discussed in both cases.

This book belongs to the *SAP Excellence* series, which initially contains the following works:

- Appelrath, Hans-Jürgen; Ritter, Jörg: SAP R/3 Implementation. Methods and Tools
- Becker, Jörg; Uhr, Wolfgang; Vering, Oliver: Retail Information Systems Based on SAP Products
- Buxmann, Peter; König, Wolfgang: Inter-organizational Cooperation with SAP Systems. Perspectives on Logistics and Service Management
- Knolmayer, Gerhard; Mertens, Peter; Zeier, Alexander: Supply Chain Management Based on SAP Systems. Order Processing in Manufacturing Companies

A feature of all these works, and thus also this book, is that employees of SAP have recently validated all statements made about the software. We have created under www.wiwi.uni-frankfurt.de/sap a discussion forum to exchange experience gained with the use of SAP systems for the cooperation between companies and, in particular, for the supply chain management. We hope that our readers actively participate in the discussion forum.

Finally, we would like to express our thanks to a number of people who supported us in many ways in the production of this book. These include Prof. Peter Mertens and Dr. Peter Zencke, the initiators of the *SAP Excellence* series, and Dr. Franz Hollich, our central contact partner at SAP AG. We also thank our partners for the practical examples, Dr. Joachim Klein and Bernhard Oymann at Schenker Logistics, Jürgen Herb at Goodyear, and many employees from SAP for their exemplary cooperation. Last but not least, Markus Fricke, Sven Grolik and Claus Hittmeyer, all employees at the Institute for Information Management, earn our grateful thanks for their valuable support in preparing notes for this book. Moreover, we would like to thank Anthony Rudd for translating the manuscript.

Peter Buxmann   Wolfgang König

Table of Contents

Chapter 1 The Use of Information and Communications Technology to Support Cooperation Between Organizations in Logistics

1.1 Forms and Motives of Building Cooperation

There are many prominent examples for inter-organizational cooperation:

- The eight airline companies Lufthansa and SAS in Europe, United Airlines in the USA, Thai Airways International in Asia, Air Canada in Canada, Varig in Brazil, Air New Zealand in New Zealand and Ansett Australia in Australia have formed a global alliance with the name "Star Alliance".

- Pharmaceutical companies cooperate in research and development projects.

- Software providers cooperate to guarantee compatibility for the customers and to build initial barriers against new competitors.

- Programmers work together in open-source projects to develop software systems. The best example of this is the development of LINUX.

There are many other examples of inter-organizational cooperation where the partners either pursue a common goal or support each others goals ("one hand washes the other").

In this book, cooperation is an implicitly or contractually agreed collaboration between independent companies. Indeed, cooperation agreements may be for a limited duration, such as for a single project. Terms such as virtual companies, virtual organizations or strategic (company) networks all describe these new forms of cooperation. Differences in the definitions between these concepts are often arbitrary, and, in our opinion, not of practical use.

The economic effects on the involved participants resulting from cooperation activities are wide-ranging. After all, they result from a clever combination of the existing resources, which can be production plants, employees or information. This combination of production factors represents the basic motivation for forming the cooperation agreements. The following benefits can be achieved:

- The classic benefit of cooperation for the involved partners is *cost reduction* that can be realized, in particular, from economies of scale and economies of scope (Boyes/Melvin 1994; Gravelle/Rees 1992).

- The *time* factor is an important reason for establishing cooperation agreements. The term "*time to market*" designates the interval from the idea or vision through to the market introduction. Many empirical investigations indicate that the life cycle of products is becoming increasingly shorter and there is a statistically significant relationship between the time of entry to the market and the market share. For this reason, cooperation agreements become attractive because the involved partners, for example, can perform tasks in parallel, which leads to reduced development times.

- The *reduction of risks* has also been frequently mentioned as being a motive for the cooperation between companies. A division of effort can also share the risk of failure. This applies, for example, to projects in research and development, in which there is normally a large potential risk.

- *Quality advantages* through cooperation can, for example, permit alliances of airlines, car rental companies and hotels to provide additional services, such as a coordinated availability and return rental cars and the crediting of bonus points.

- Cooperation can also result in an increased *flexibility* by permitting access to additional production capacity of the partner (flexible capacity expansion).

- In particular, the inclusion of employees and their knowledge can result in an increased *innovation activity*. New product innovations may differentiate the product offerings and provide additional competitive advantage.

- Cooperation can simplify *access to new markets* and remove traditional industry boundaries. Or, cooperation can provide a high market share, which then increases the barriers to market entry. This permits the cooperating firms to charge higher prices in the market.

1.2   Inter-organizational Cooperation in Logistics

In this book we concentrate on inter-organizational cooperation in logistics. Thus, the basic principle of Supply Chain Management (refer to Section 2.3) considers the participants of a logistics chain to be partners that cooperate with each other in order to better satisfy the requirements of the end-users. The inter-organizational optimization of the goods and information flows assumes primary importance here. This means that the partners become closer and thus the classic company boundaries become increasingly blurred. Examples of such close cooperation in the logistics are:

- As part of their partnerships, the suppliers and customers work together to create products. This cooperation starts with the first concept, includes process development and continues until the market introduction of the product.

- Suppliers are compensated based on the commercial success of the producer. An example here is the development of the Smart Car (also refer to Section 2.3.3).

- Manufacturers use information to determine the supply quantity and delivery dates to ensure demand-oriented supply to the regional and central warehouse, and the sales locations.

- Schenker Logistics, a logistics service organization, cooperates with Mercedes-Benz on the management of a supply chain for the manufacturing of the A Class model (refer to Chapter 7). The company is primarily involved in the area of IT-usage of the supply chain.

Integration is increasing because of global competition and progress in information and communication technology. The hypothesis: the efficient use of information and communication technology, and the organizational willingness to use networks, will play a key role in control of the information and goods flows between the involved partners. In this book, we primarily discuss the use of the new technologies.

1.3   Development Stages in Information and Communication Technology

When we consider both the development stages in the information and communication technology and company logistics, it becomes clear that the trend in both cases has changed from internal support to inter-organizational cooperation.

Figure 1.1 shows the development of the use of information and communication technology over previous decades.

In the 1960s, companies increasingly began to use information and communication technology to support individual areas. Initially, these were often the accounting and manufacturing areas.

In the next phase, support was also provided to cross-functional areas. This resulted in an integration of different functions, such as cost accounting with financial accounting.

Since the mid-1980s, software has increasingly focused on business processes. The best example here are the SAP products for which a business process orientation has increasingly replaced the traditional functional orientation.

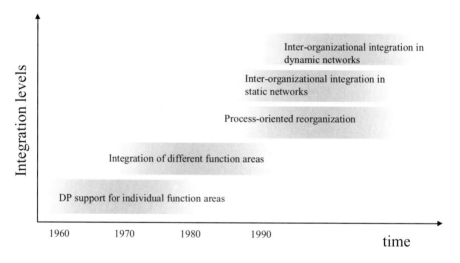

Figure 1.1: Developments in the information and communication technology

Since the mid-1980s, the support for inter-organizational integration has been a new trend in information and communication technology. Ten years later, the Internet intensified this development and increased the growth rates. Terms such as electronic commerce, Intranet, Extranet or "Electronic Data Interchange (EDI) over the Internet" show the relevance of business processes on networks. Providers of standard business software, such as Baan, J. D. Edwards, Oracle, Peoplesoft and SAP, pursue these developments and provide solutions to support inter-organizational cooperation.

Currently, this inter-organizational integration is normally realized using static networks. For example, such close relationships between producers and suppliers can often be found in the automobile industry. These are long-term relationships between companies that support the appropriate technologies, for example EDI. The investments needed to establish these relationships normally mean that a flexible integration of other suppliers can either not be realized or only at high cost.

Organizations increasingly cooperate on the basis of short-term relationships. The vision is currently met in just a few cases: the cooperating partners provide relatively specialized solutions, which they offer over standardized interfaces on the network and that can be readily replaced. The prerequisite here is the availability of interfaces at the technical and logical level to ensure a high degree of flexibility at relatively low cost.

The economical advantage of the use of information and communication technology is primarily in the reduction of the information and communication technology costs between the cooperating partners. The following example for 3COM illustrates this relationship.

With more than six billion dollar sales, 3COM is worldwide one of the leading companies in the area of electronics and network communication. Mainly at the request of a large American business partner, 3COM implemented in 1995 an EDI solution based on the ANSI X12 Standard. Thanks to the continuing globalization and the international business partners using this standard, EDIFACT was also introduced as EDI Standard in 1998. 3COM uses EDI with approximately 15% of its approximately 200 business partners (customers and suppliers). The restructuring of the distribution channel has already started to concentrate on the larger suppliers, which in all probability are EDI users. This also explains the fact that approximately 50% of the suppliers already support EDI. Because EDI is becoming increasingly a strategic factor in the computer industry, 3COM endeavors to convince its business partner to use this technology. EDI capability is already a necessity for new suppliers and intermediaries. It can be expected that financial service providers will also be integrated in the 3COM EDI network to process payments in the course of the next year.

As in many other companies, the EDI solution from 3COM was originally based on PC technology. However, because of the large increase in the data volumes being transferred, but also, in particular, forced through the acquisition of U. S. Robotics, the solution has also been migrated to a UNIX platform. 3COM uses a private Value Added Network (VAN) from IBM for the data transfer.

When we consider the cost of the implementation, we can differentiate between four areas:

- The one-off cost for the EDI solution was approximately $ 25,000 (including the VAN costs for the first year). These costs were relatively low because existing technical and personal resources could be used.
- The increasing data traffic required an *extension to the system*. Additional personnel were employed and the EDI Operations Department was formed. The improvement of the technical infrastructure required $ 100,000 for a new UNIX conversion program.
- Two to three programmer-days are required at 3COM to *connect a new business partner* to the existing system.
- *The integration of a new transaction/document type* into the existing EDI set requires approximately eight man-days; the equipment costs are approximately $ 1,140.

The running costs are $ 350,000 for personnel, $ 36,000 for the data transmission (VAN) and approximately $ 17,000 for additional external services, such as consulting.

If we now consider the specific benefits of the EDI application: 3COM estimates that the cost for processing a document has fallen from $ 38 dollar to $ 1.35. This is a total annual savings of $ 750,000 for the ordering and invoicing processes. Together with the savings from reduced error rates for the data input, improved inventory management and faster process execution, there is a total savings of $ 1.3 million.

The acquisition of U. S. Robotics in mid-1999, meant that 3COM was in the consolidation phase. This, and the concurrent integration of the two EDI systems of both two companies, means that the savings are expected to increase in year 2000. 3COM plans to use the Internet as transport medium in the future. In addition to form-based EDI, the integration of the various computer systems at both parties should also be realized. The use of XML is also being considered here. 3COM expects that the use of the Internet will reduce transmission costs, and the response times, compared with traditional batch processes in VANs, should also be reduced considerably.

1.4   Development Phases in Logistics

The logistics area shows parallels to the previously discussed development in the area of information and communications technology. At the start of the 1970s, and thus roughly at the time of appearance of the first German-language contributions to "business logistics", the various subareas – production, distribution and sales – operated relatively independent of each other. The attempt made to optimize the logistics subfunctions using EDI had two effects. On the one hand, the expected improvements in the individual areas occurred and inefficiencies were reduced. On the other hand, it became clear that the optimization of the subfunctions did not produce an overall optimization. Indeed, it proved to be counterproductive from a company-wide viewpoint.

In the 1980s, the business process departed from the function-oriented concept and behavior methodology. The new concept of the company as "flow system" arose. The potential advantages of centrally coordinating the company-wide logistics were now recognized. This permitted an improvement in the customer benefits coupled with a simultaneous reduction of the logistics costs, including the cost of the capital tied up in the warehouse. The consequence of these new concepts and behaviors was a linking and integration of the logistics subfunctions and the interconnection of existing stand-alone solutions.

Recently, the integration efforts were extended to the inter-organizational level. As also seen in the area of the information and communications technology, networks and strategic enterprise alliances resulted. Independent enterprise units cooperated to solve marketing and production tasks together.

As previously shown in this chapter, the processes of networking and the integration occur at various levels. Somewhat simplified, this process contains both an organizational and a technical dimension. These two components determine the network capability of the participants. If the appropriate capabilities are available, their chances to succeed in a global world economy increase. The capability for networking will become a competitive factor.

Chapter 2   Inter-Organizational Processes and Cooperation in Logistics Networks

Material procurement for production and distribution of consumer goods results in raw materials, semi-finished goods and finished goods passing through various stations in a logistics network. In particular, these are:

- suppliers
- producers
- wholesalers
- transport companies
- warehouses and goods distribution centers
- logistics service providers
- retail branches.

In classic business studies and logistics management methods, the participants in a logistics network are considered isolated without having any system connection. It is assumed that each individual participant himself decides on the economics for the procurement, provision of services and products, and sales, without aiming to achieve a coordination in the complete network.

Only in the mid-1980s with Hulihahn's work was the network as an entity studied under the term supply chain (Hulihahn 1985). The consideration of the supply chain as a unit aims at coordinating the goods and information flows in the complete network, and so increase the value for the end-customer. In addition to the goods and information flows as input and output of value processes, financial flows, for example, must also be taken into consideration. Such a coordination is also known as supply chain management and is based on the cooperation of all participants in the logistics network.

The following discussions initially explain the basic logistics activities and methods. Section 2.1 discusses the materials logistics, which as part of the provision of logistics services, covers in particular transport, transshipment and storage of physical goods. Section 2.2 then describes the information logistics, which embodies that part of the logistics concerned with the planning and control of information flows and the transport of the information in a company and its environment. In Section 2.3 we use these discussions as basis for the representation of the supply chain management as strategic management concept for the planning, control, realization and supervision of the goods and information flows in a logistics network and along the complete supply chain. We then describe in Section 2.4 modern task areas for logistics service companies in the supply chain and pay special attention to the role of logistics service providers as supply chain managers. To close the chapter, Section 2.5 provides a short summary of the supply chain management systems currently available.

2.1   Principles of Materials Logistics

The aim of the materials logistics is to ensure the availability of the physical goods – such as raw materials, semifinished goods and finished goods – when needed in logistics chains. The materials logistics has the task to provide here
- the right goods
- in the right quantity
- in the right quality
- at the right location
- at the right time, and
- at the right cost

in accordance with an appropriate agreement.

Various central services for logistics, such as transport, transshipment and storage, but also logistics supplementary services, such as picking and packing, must be provided here. Furthermore, many services that arise in conjunction with logistics services must be taken into consideration as additional supplementary services. These supplementary services represent a high potential for integration. An example are the quality tests for the transported goods that a logistics company performs on behalf of the supplier or the customer, or both.

All activities to ensure the appropriate availability of the physical goods must focus on the specified goals. These include the optimization of the logistics service with regard to service criteria such as delivery time and reliability of delivery, but also for cost types such as transport and storage costs. Furthermore, ecological criteria must also be taken into consideration. For example, a minimization of the emissions of carbon dioxide and sulfur oxide may be required for the provided transport services.

The materials logistics can be organized into functional subsystems for the procurement logistics, the production logistics, the distribution logistics and the reverse logistics. Considering the background of the importance for the inter-organizational coordination of the goods flow, the following Section 2.1.1 discusses the functions for procurement logistics and distribution logistics in more detail. Finally, Section 2.1.2 discusses transport chains and transport networks as basic planning units for the materials logistics. Section 2.1.3 then follows with a description of the various planning and optimization concepts to define goods flows as part of the materials logistics.

2.1.1 Subfunctions for Procurement Logistics and Distribution Logistics

Figure 2.1 shows a schematic simplification of the procurement logistics and distribution logistics.

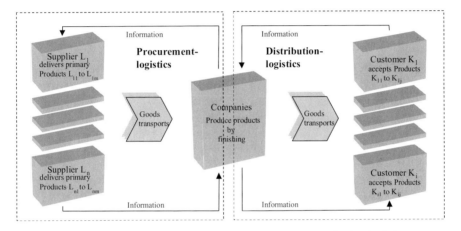

Figure 2.1: Transport in the procurement logistics and distribution logistics

The discussion here centers on a company that procures goods, further processes these and sells the produced goods to customers. For example, a tire manufacturer can obtain raw materials from suppliers, produce tires from these in a plant, and sell the tires as products to car companies. The goods transport are in the reverse direction with regard to the information transport (information flows), namely the orders from the company with regard to the companies. The transport flows shown in Figure 2.1 can also be found within companies, when, for example, there is the need to transport intermediate products from various workplaces to an assembly hall and products to various departments for further processing.

The procurement logistics is responsible for the planning, control and implementation of all required structures and processes to provide the appropriate supply of the company with goods. The procurement logistics covers various subfunctions and coordinates them under a common destination system, in particular with regard to performance and cost aims. The subfunctions include the determination of the requirements, assignment of the deliveries, selection of the suppliers, determining the time of ordering, materials scheduling, transport to the plant, goods acceptance, testing the goods, container handling, warehousing, storing, sorting, picking, marking, transport within the company, and provision of materials for production and other company areas.

The distribution logistics cover all physical, scheduling and administrative processes concerned with the distribution of goods from an industrial or trading company to the subsequent business level or to the consumers. After making allowance for the requirements of the delivery service, as discussed previously, it remains to transport the right goods in the right quality and quantity at the right time to the right location. The business aim is to bridge this space and time as economically as possible. Subfunctions of the distribution logistics are both long-term tasks, such as the planning of the transport network (refer to Section 2.1.3) and in this connection, the selection of the location of the distribution warehouses, the planning and layout of the warehouses, and also short-term tasks, such as storage, the picking and packing, the vehicle routing with the associated planning of transport resources and delivery dates, and the order processing. Because the customers are increasing their demands on the delivery service, the modern distribution services must provide additional capabilities other than just the logistics services. For example, to reduce their own costs, many customers demand certain services from the supplier, such as price labeling of the products.

Figure 2.2 shows how trading companies frequently handle the distribution of goods to the end-customers.

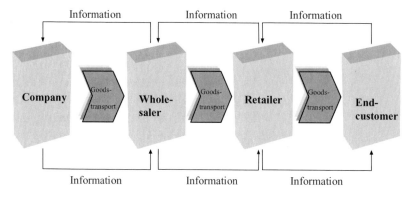

Figure 2.2: Wholesalers and retailers connecting a production company to the end-customers

Such a trading organization permits, for example, the production of cars to be concentrated at increasingly fewer locations in the world, while the delivery, and in particular, the service associated with automobiles can be provided decentrally over a wide area.

2.1.2 Transport Chains and Transport Networks

Materials logistics are primarily concerned with the transport chain. In accordance with DIN 30781, a transport chain is a sequence of linked technical and organizational activities used to transport goods from a source (starting location) to a sink (destination). The processes involved in the transport chain do not concern just the location changes of the considered objects, but actually to all transformation processes between a source and a sink. In addition to the transport, these can be processes such as packaging, storage, transshipment or customs clearance.

Figure 2.3 describes a typical scenario for a transport between a consignor (supplier) and a recipient ((end-)customer). This is a goods transport chain in which a supplier uses a contracted forwarding agent to ship specific goods to a customer (pre-carriage). In addition to this local transport, the transport company undertakes the complete organization of the transport along the shown (sub)chain. This focuses on the selection of the carrier for the physical transfer over long distances (road, railroad, air, sea shipping companies) to the appropriate receiving agent. The forwarding agent (or receiving agent) or the transport company performs these transshipment processes. The receiving agent delivers the goods to the recipient as the post-carriage activity.

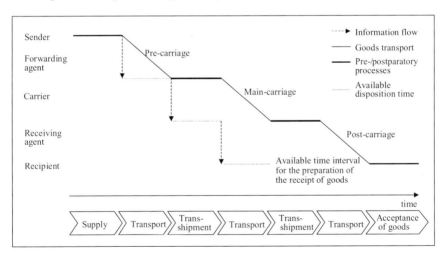

Figure 2.3: Involvement in goods transport chains (Wolf 1997, page 1093)

The goods flow between several specified supply and receiving points bridges space and time in transport networks. In single-level transport networks, this goods flow progresses directly from the supply points to the receiving points. The principal advantage here is the uninterrupted goods flow, which, for example, does not require any intermediate storage processes. In contrast, this bridging in multi-level networks takes place as an indirect goods flow between the supply points and the receiving points. Figure 2.4 emphasizes these interruptions as a dissolution or concentration point. The aim here is, for example, to reduce the transport costs by combining transports. A number of different transport networks, such as the constellation of a combined network, is always conceivable, which occurs when direct and indirect goods flows are combined with each other.

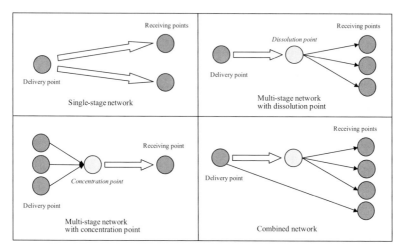

Figure 2.4: Basic structures of transport networks (Pfohl 1990)

2.1.3   Planning and Optimization Concepts in Materials Logistics

This section discusses several planning and optimization concepts used in the procurement and distribution logistics. Common to all the following models is that you must set the appropriate parameters to use them or access the associated data in the associated databases. The methods are also provided in standard software adapted to customer requirements or based on customized software.

In this section we restrict ourselves to short discussions of simple planning and optimization concepts. We will often present static models, namely methods in which parameter values do not change during the planning period, and, when appropriate, provide references to more advanced literature for associated dynamic models.

Supplier assessment and selection

The need for goods in the company produces the problem of finding and selecting suppliers. Scoring models can be used to select and assess a supplier:

The significant criteria for the assessment must be formulated in an initial step. These are such things as the price for a commodity to be supplied, its quality, the geographical distance from the supplier, the flexibility with regard to changing supplier requirements, and its reliability. The criteria are weighted once the criteria catalog has been defined.

The actual assessment of the suppliers is performed in the next step; a value is assigned to each individual criterion representing the extent to which it is satisfied. A value within the range 1 to 10 points can be assigned as degree of fulfillment in the following example (a large value represents high fulfillment). The assessment is made as a matrix such as that shown in Figure 2.5.

Criteria (weigh-ting) Sup-plier	Commodity price (0,35)	Com-modity quality (0,2)	Geographical distance (0,15)	Flexibility (0,1)	Reliability (0,2)	Total (1,0)
Supplier A	4	2	6	5	3	3,8
Supplier B	8	5	4	9	5	6,3
Supplier C	4	7	8	2	5	5,2

Figure 2.5: Assessment for three suppliers

The total value for a supplier is determined in two steps:

- through the multiplication of the charge factor value for each criterion with the defined weighting of the considered criterion, and
- the subsequent addition of all criteria values.

Finally, that supplier with the highest total value can be selected (this is supplier B in Figure 2.5).

When scoring models are used, note that a subjective procedure characterizes the selection of the evaluation criteria and the weighting factors, and the assessment of the suppliers using the individual evaluation criteria. Furthermore, no consideration is made of any dependencies between the criteria.

ABC analysis

The ABC analysis is an analysis method used, for example, to differentiate between goods, customers and suppliers. We illustrate the ABC analysis in the area of supply politics using the example of the differentiation of the quantity of parts to be procured in the companies. The procedure divides the range of parts required by a company into three classes (for reasons of practicability). The range of parts to be procured is classified using a defined value criterion, such as demand value or inventory value (each per time unit). The aim is to determine a value-quantity factor that reflects the relative importance of a part type.

The ABC analysis for an assignment of the parts, such as using the criterion of the annual demand value, proceeds in the following steps:

- First determine the annual consumption quantity for each part and, in the simple case, the average procurement price per unit.
- Then, through multiplication, determine the annual consumption value for each part and sort the range of parts into decreasing sequence.
- In the decreasing sequence of the annual consumption value, calculate the percentage factor of the total annual consumption value and of the consumption quantity for each individual parts position.
- Then, also in decreasing sequence, enter the accumulated results into a diagram as shown in Figure 2.6 and define the class limits by setting a limit at two specific percentage factors of the total demand quantity.

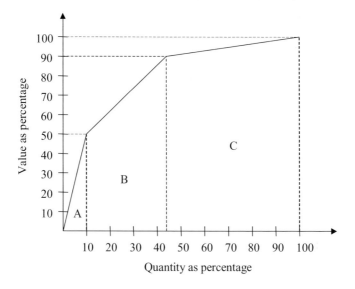

Figure 2.6: Principle representation of the ABC analysis (Schulte 1995, page 162)

This normally produces a Lorence curve that can have various distinctive forms. For example, the in case shown in Figure 2.6, only ten percent of sales by quantity represent 50 percent of sales by value. This indicates that these parts, the so-called A-parts, should be ordered using a more exact, and thus more comprehensive, planning and optimization procedure, whereas the C-parts, which represent higher quantity sales but lower value sales, should tend to ordered using simple inventory models.

In addition to the analysis of the relative importance of a part type during the ABC analysis, the continuity or discontinuity of consumption is also important for the identification of the most suitable procurement form. For this reason, a so-called XYZ analysis is frequently performed to augment the ABC analysis; this analyzes the material items with regard to their consumption structure (whereas X-parts show a very stabile consumption structure, that of Z-parts is very irregular).

Inventory and stock ordering policies

Warehouses serve to compensate for the different speeds of production or delivery and the acceptance. The various inventory models prepared in economics make different assumptions with regard to the planning situation and thus represent many possible problem situations. In particular, a differentiation can be made between static and dynamic inventory models. In contrast to the static concepts, dynamic models explicitly handle the time aspect. In particular, they cater for the fact that the material requirement varies during the planning period. We restrict the following discussions to the representation of the simple static inventory model. Generally, support must be provided for two types of ordering decisions:

- When should an order be made to fill a warehouse? The order point or reorder level here is that inventory level which when undershot suggests or automatically initiates a reorder.
- Which quantity (lot size) should be ordered?

Use the following stylized change of the inventory quantity for each item to determine the reorder level (refer to Figure 2.7):

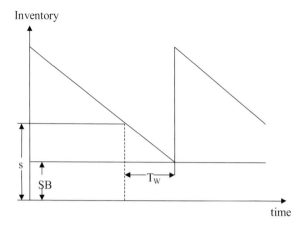

Figure 2.7: Typical change in the inventory level

Starting with an initial inventory level, the inventory reduces with a constant depletion rate B_t (for example, average inventory depletion per day). The time procedure tests for each individual stock withdrawal whether the reorder level has now been undershot. The reorder level s for each inventory item is calculated as follows:

$$s = B_t \cdot T_w + SB$$

Multiply the lead time T_w (days in this example) with the average stock withdrawals per day B_t and add the safety stock SB.

To determine the optimum ordering quantity, we restrict ourselves to the representation of the simple static Andler lot size formula. This assumes that the total requirement quantity B is known for a planning period (for example, one year). We also use constant warehousing cost unit rates 1 for the goods and also constant fixed costs for each procurement activity K_f. The procurement costs in the planning period K_B depend on the number n of procurement activities and yield

$$K_B = n \cdot K_f = B/q \cdot K_f$$

where q represents the (constant) ordering quantity for each order. The K_B reduces with increasing q; it has hyperbolic form shown in Figure 2.8:

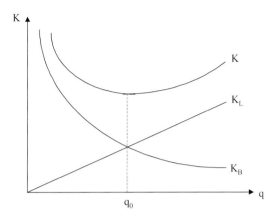

Figure 2.8: The optimum quantity in the classic case (Ihde 1991, page 207)

Assuming that the inventory level per period is depleted n times by the amount q to zero, on average exactly a half lot is always available in the warehouse. If we assume the warehousing cost unit rate 1 for this inventory, the period-related warehousing cost K_L is then:

$$K_L = 1 \cdot q/2$$

The warehousing cost unit rate contains the unit and period-related costs K for the warehousing and the tie-up of capital. Because the cost types K_B and K_L change inversely and continuously, an optimum ordering quantity can be determined that minimizes the total cost

$$K = B/q \cdot K_f + 1 \cdot q/2$$

Differentiate (according to q) and set to zero to provide the classic lot size formula:

$$q_0 = \sqrt{\frac{2 \cdot B \cdot K_f}{1}}$$

To make the Andler model more usable in practice, it has been extended in many ways, for example to cater for the finite inventory replenishment rate and to include (quantity) discounts (refer, for example, to Hadley/Whitin 1963) and shortage.

Location planning

The site location planning adopts a major importance in the distribution logistics. As example, we consider the location planning for distribution warehouses. There are a large number of possible planning and optimization concepts. The following parameters, for example, can be used to differentiate between these concepts: the number of distribution levels to be considered (one-level versus multi-level distribution), size of potential locations (with or without capacity limits), and the degree of freedom for the geographic placement of locations (select from a set of prespecified potential locations versus "free" (unrestricted) placement of an optimum distribution warehouse).

In the following section, we first handle a (simple) single-level selection problem from a set of prespecified possible locations. This is the so-called Warehouse Location Problem subject to capacity limits. We start with the following premises (also refer to Figure 2.9): A company with a single product must supply n customers that have a demand of b_j quantity units (j=1,...,n) per period. The company wishes to reduce its distribution costs by building distribution warehouses. The basic idea is that the collective transport from producers to the distribution warehouses bring cost advantages compared with a direct delivery to each customer. A finite number of potential warehouse locations i=1,...,m with capacity a_i are available for building the distribution warehouses. If a warehouse is built at the potential location, it has i has fixed costs of f_i per period. The transport costs between each warehouse location i and a customer j are assumed to be proportional to the transport volume; these are c_{ij} money units per transported quantity unit. This quantity can also include the variable warehouse costs at location i. The transport costs from the producer to the distribution warehouses are ignored. Because these costs are similar for a preselection of specific potential warehouse locations in a region and thus not relevant for the decision in our problem, we are justified in neglecting them.

The optimization task is to determine those locations from the set of the potential locations (and to build warehouses there) that minimize (catering for the secondary condition of the complete delivery of all customers) the total for the warehouse and transport costs. We introduce two decision variables here: y_i assumes the value 1 when a warehouse is to be built at the potential location i, otherwise the variable has the value zero. The decision variable x_{ij} represents the number of product units to be supplied from location i to the requester j.

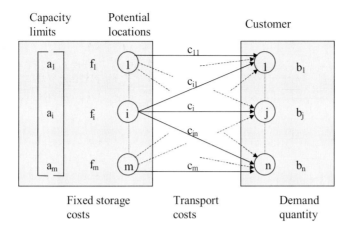

Figure 2.9: Single-level Warehouse Location Problem subject to
capacity limits (Domschke/Drexl 1996, page 52)

The required function:

$$\text{minimize } K(x,y) = \sum_{i=1}^{m} f_i y_i + \sum_{i=1}^{m} \sum_{j=1}^{n} c_{ij} x_{ij}$$

The following restrictions apply:

$$\sum_{i=1}^{m} x_{ij} = b_j \qquad \text{for } j=1,...,n \qquad \text{(all customer demands are satisfied)}$$

$$\sum_{j=1}^{n} x_{ij} \leq a_i y_i \qquad \text{for } i=1,...,m \qquad \text{(the capacity restrictions of the distribution warehouse are observed)}$$

$y_i \in \{0,1\}$ and $x_{ij} \geq 0$ for $i=1,...,m$ and $j=1,...,n$

We also wish to present the Steiner-Weber model as a representative of non-discrete location planning. We make the following assumptions here (also refer to Figure 2.10): a company with a single product is looking for a distribution warehouse location from which it can supply its customers at minimum cost. To simplify the calculations, an unlimited homogeneous area is assumed that has n customers at the coordinates (u_j, v_j) each with supply requirement of b_j quantity units (for all $j=1,...,n$). Each point of the area is a potential location for the warehouse. The transport costs between any two points of the area are assumed to be proportional to the distance and the transport volume. The cost per distance unit and transported quantity unit is c money units.

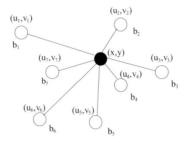

Figure 2.10: Possible network structure for the free placement of a distribution warehouse

The Euclidean distance (namely the shortest distance) is assumed to be the distance between any two points. The costs for the transport between production plant(s) and warehouses are neglected.

The optimization task is to find those coordinates (x,y) for the distribution warehouses for which the customers can be supplied at minimum cost. The required function:

$$\text{minimize } K(x,y)= \quad c \cdot \sum_{j=1}^{n} b_j \cdot \sqrt{(x - u_j)^2 + (y - v_j)^2}$$

The Miehle iteration method can be used to solve this problem.

Vehicle routing and scheduling

The standard problem for the vehicle planning and scheduling involves using a vehicle from a depot (for example, distribution warehouse) to supply the required goods to a specified number of customers for which the quantities and locations are known. No partial deliveries are permitted; each customer is visited just once. The optimization task is to plan the route (journey) so that the total traveled distance is minimized while satisfying the customer requirements and observing any specified restrictions (for example, capacity and time restrictions). The vehicle must be back at the depot at the end of the journey.

The (standard) vehicle planning and scheduling problems involve the customers being assigned to various stretches and determining the sequence of the visits to the customers. Exact and heuristic solution methods exist to solve such vehicle planning and scheduling problems. Because the solution of realistically large problem situations (with regard to the number of customers) using exact procedures is not normally currently possible with an economically viable effort, the following section describes a heuristic procedure, the Savings procedure. This method is used in various software products.

Figure 2.11 illustrates the principle of the method. The depot is indicated as location 0; the numbers 1, ..., n (here 5) indicate the customers. We start with an initial solution in which each individual customer i is served alone from the depot, thus the vehicle travels from the depot to a customer and then returns directly to the depot. In the following phase of improvement over the initial situation, we attempt incrementally to improve on this initial solution by combining two neighboring trips provided this does not conflict with the capacity or time restrictions. For example, the vehicle no longer returns from customer 2 to the depot but goes directly to customer 4 and only then back to the depot. If we designate the first and the last customer of a route as end-customers, two trips are combined through the transition from an end-customer of the route of the first to an end-customer of the route of the second trip.

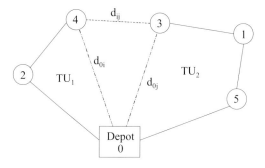

Figure 2.11: Combining of trips or routes

The combining of two trips by joining their routes over the end-customers i and j yields a saving s_{ij} of the amount

$$s_{ij} = d_{0i} + d_{0j} - d_{ij}$$

where d_{0i} and d_{0j} represent the distance from customers i and j from the depot and d_{ij} represents the distance between the two customers. s_{ij} becomes larger the nearer the customers i and j lie together and the further distant they are from the depot. Every iteration of the Savings procedure takes account of the restrictions to realize the linking of those two trips that provide the largest savings. The procedure ends when no further combining of two original trips can be included in the solution.

Alternative improvement methods are the 2-opt and the 3-opt procedures. These procedures are based on an initial solution with a systematic exchange of edges. In the 2-opt procedure, two edges, such as a journey that connects the towns 1 and 2, and 3 and 4, are replaced, for example by joining the towns 1 and 3, and 2 and 4. Such an exchange is performed and processing continued with the solution in the next iterations step when this produces a reduction in the journey. In the 3-opt procedure, three edges of a journey are systematically replaced with three other edges. Such heuristic solution methods do not offer any guarantee that the optimum is achieved.

Storage space planning (Packing)

The storage space planning aims to provide the best possible usage of storage space capacities for loading aid and transport resources, such as the loading of pallets, containers or trucks.

For homogeneous storage space problems, so-called congruent packing items, such as cuboid packages or parts with a circular base (such as cans, bottles or barrels) are to be arranged to maximize the number of included packing items.

To illustrate the procedure, the following section analyzes the simplified case in which the height h of a packing item does not play any role in the planning with regard to the height H of the available storage space (for example, a pallet is loaded only with one "layer" of parts that can have different heights, but which are less than H; l, b and h, and L, B and H are each perpendicular to each other). On the two-dimensional base, for example a pallet (with length L and width B), the maximum number x_{max} of rectangles with the dimensions $l \cdot b$ are to be arranged orthogonally. An optimum layer plan uses selected coordinates of the packing items or a graphical representation to describe how the packing items are to be positioned on the base of the loading aid (refer to Figure 2.12).

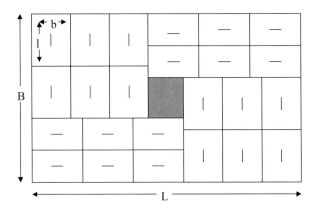

Figure 2.12: Layer plan with 4-block structure (Isermann 1998b, page 17)

When the use of exact solution methods, such as the integer optimization, permits optimum layer plans to be generated, heuristic procedures are frequently used to solve practical problem situations. The common heuristic procedures assume a division of the pallet base into rectangular sub-bases (so-called blocks) and assign the sub-bases that are not further subdivided with the maximum possible number of packing items either diagonally (the b-side of the packing item is arranged parallel to the L-side of the base) or lengthwise.

Within the class of the heterogeneous storage space problems, such as in a cuboid storage space with length L, width B and height H, the maximum possible number of parts from a list of packing items $P_1,...,P_N$ are to be included. Such items, for example, are arranged according to volume or weight of the packing items or other decisive criteria. Additional restrictions can apply, such as for a truck loading area that is accessible only from the back. Other examples occur in the delivery sequence specified for vehicle routing for batched delivery contracts and prohibited loading combinations for dangerous goods. However, even the center of gravity or the three-dimensional weight distribution can be significant. The practical aims here are, for example, the minimization of the number of loading aid items, the minimization of the unused storage space or the minimization of the transport and handling costs. Heuristic procedures attempt, for example, a layerwise arrangement of identical or similarly dimensioned packing items (thus a transformation of this problem situation to the solution of a homogeneous storage space problem) subject to specified restrictions (for example, position of the center of gravity, weight distribution, prohibited loading combination). Refer to Gehring, etc. (1990), for additional heuristic concepts for heterogeneous storage space planning.

2.2   Principles of the Information Logistics

The information logistics represents that part of logistics concerned with the planning and control of information flows and the transport of information in a company and its environment. This concentrates on the information used for the planning and control of

*   goods flows
*   financial flows
*   the assignment of people who provide services.

The aim of the information flow is to ensure an optimum coordination of flows for real and nominal goods.

This background makes it clear that information logistics requires a complete and agreed planning, structure and use of company-internal and external information systems, and thus information systems that are directly accessible using interfaces. The goal and, at the same time, challenge for management is the realization of the *"logistics principle of information"*. This requires the availability of the right information at the right time at the right location in the needed quality and in an appropriately detailed and correct state. These can provide the prerequisites for information to replace the losses that occur at interfaces between the various members of a value chain, such as those caused by large data quantities.

A central goal of the information logistics concerns the optimization of the information availability and the information throughput times. This results in methods such as the Just-in-Time (JiT) principle, which says that production factors should be made available only when they are actually needed for the production. This principle is used, for example, in the automobile industry, where suppliers provide the ordered goods within a short time after a request for delivery. In analogy to the physical level, because the transport of information is subject to logistics considerations in the same way as the transport of real goods, the JiT concept can also be applied to the processing of information.

The next section initially describes how information logistics and materials logistics interact with each other (Section 2.2.1). Then, because of their importance for the definition of organization forms for logistics networks, we briefly present several important modern information and communications technologies and consider, in particular, the Internet as the modern communications backbone (Section 2.2.2). Section 2.2.3 discusses the importance of the information and communications technologies in conjunction with Electronic Commerce and differentiate here between the individual phases of a business relationship, namely the initiation, negotiation and execution phases. Section 2.2.4 then describes the security in information and communications networks, in which a differentiation is made between closed and public networks. Finally, to complete these discussions on the principles of information logistics, Section 2.2.5 describes the role of standards and the closely related topic on the importance of network effects.

2.2.1   The Connection between Information and Materials Logistics

In materials logistics, the planning, control and monitoring of the provision of logistics services is based on information. The planning, control and monitoring of the information flows is performed here as part of the information logistics. Information and materials logistics affect identical processes and areas within and between companies, and make use of each other (refer to Figure 2.13).

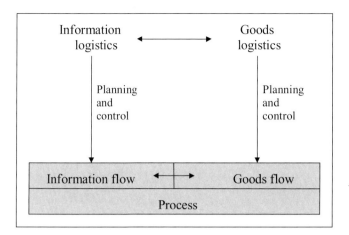

Figure 2.13: The interaction between information and materials logistics

The following discussions emphasize the effect the information logistics have on the materials logistics through the control of the information flows:

Information directly concerned with the definition and control of the goods flow can differ between that required for ordering, planning and reservation, and that required for transport and picking.

Ordering, planning and reservation information is used to prepare the goods flow. Such an information flow can be controlled so that it either runs counter to the goods flow or initiates from a central instance (refer to Figure 2.14). The first planning and control policy is the so-called Pull Principle. For example, ordering information is sent from manufacturing location 2 to a forward manufacturing location 1, which there initiates an appropriate goods flow to the manufacturing location 2 (Lackes 1996, pages 840-841; Adam 1990, page 813). The same principle is used in the logistics chain when a producer orders goods from a supplier. Chapter 6 presents an application example for this situation. When the information or control impulse does not proceed in the opposite direction to the goods flow, but is sent from a central planning and control instance to all relevant manufacturing locations, this is called the Push Principle, which, for example, is realized in the Material Requirements Planning (MRP) concepts (Lackes 1996, pages 840-842).

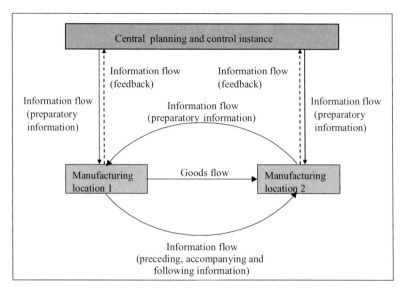

Figure 2.14: Flow directions for information and goods

In contrast to the prepared information, the transport and consignment information runs parallel to the goods flow. The information flow here can be controlled so that it precedes, accompanies or follows the goods flow. Figure 2.15 shows a comparison of the two cases of following information and preceding information. The second case here shows that the elimination of waiting times for information and a parallelization of the both the information acquisition and processing, and the goods flow during the transport process, can produce a large reduction of the throughput time for goods.

This becomes evident in the case of the transport of products to a distribution warehouse in which the products are picked and then sent to different customers. If the corresponding transport and consignment information arrives at the distribution warehouse before the products, the information acquisition and processing of the transport and consignment data for the order-picking, such as for a vehicle routing, can take place in parallel to the transport of the products to the warehouse. This means that the physical order-picking process can start immediately after the arrival of the products. This can significantly reduce the throughput time of the goods, in particular compared with the case in which the transport and consignment information arrive after the associated goods. This makes it evident that the form of the information flow, characterized here with the time of the arrival of the information, has a significant effect on the planning, control and form of the goods flow, and thus on the materials logistics.

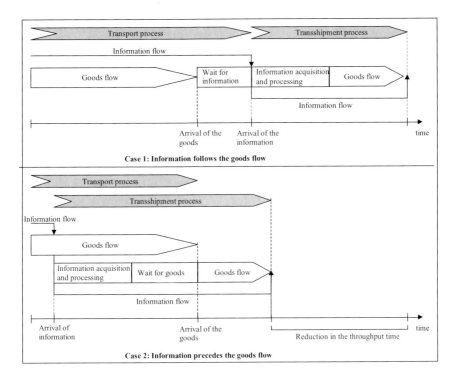

Figure 2.15: Effects of the information flow form on the materials logistics

On the other hand, because the information required for the provision of logistics services must be made available in such a manner to ensure an optimum provision of services for the materials logistics, the materials logistics also affect the information logistics.

The complete logistics planning, control and monitoring tasks are often subdivided into a logistics information system and a logistics processing system (refer to Figure 2.16). Whereas the processing system is concerned with the physical handling of the goods, the information system has the task of providing the associated information.

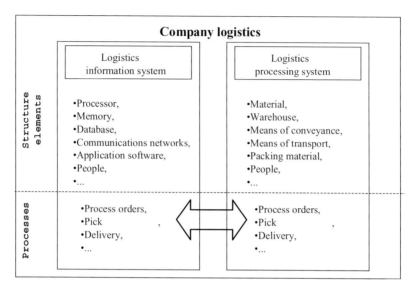

Figure 2.16: (Sub)Systems for the company logistics (Krieger 1995, page 10)

The structure of the two (sub)systems differ only in the included structure elements. In contrast, the processes associated with both systems are the same.

The interrelationships shown between information and materials logistics require a common consideration of the information and goods flows for a complete system and thus an integrated structure of the information and processing systems (Krieger 1995, pages 9-11).

2.2.2   Information-logistical Infrastructure

The basis for information flows is the existence of an information-logistical infrastructure that provides the prerequisites for the efficient processing and coordination of business processes. Modern information and communications technologies are increasingly gaining in importance over conventional media such as post, fax and telephone. The progress made in the information and communications technology also supports the structural adaptation of companies to meet the changing competitive conditions. This means, in particular, the development of new models for task distribution between companies.

Considering its importance for the structure of coordination forms in logistics chains, the following section briefly describes some important information and communications technologies and concepts. This description concentrates on the representation of the Internet as driver for important concepts used to build distributed information systems. We also present the Intranet and the Extranet as information and communications networks based on the Internet protocol. Furthermore, we discuss concepts from the Computer Supported Cooperative Work research area and finally describe the role of network services for the implementation of an information-logistical infrastructure.

Internet

The Internet represents an integration of autonomous computers that exchange information using the TCP/IP Internet protocol. The Internet basis permits various standardized Internet services to be used in the same manner independent of the location. The World Wide Web (WWW) service is the basis for the enormous popularity of the Internet. The WWW designates the sum of all hypertext documents and Web-pages stored in the Internet that can be invoked and displayed using a special client, the so-called "browser". The basic technologies for the WWW are principally the HTTP (Hypertext Transfer Protocol) transmission protocol, the browser and Web-server used to communicate, and the HTML (Hypertext Markup Language) document format used to describe Web-pages.

The Internet is an important component of the information-logistical infrastructure. Because of the low costs for the connection and transmission, it is simple to offer and request goods, information or services over the Internet. An important prerequisite here is the capability to integrate the WWW application and traditional information processing technologies. Thus, a Web interface can be used to connect traditional database systems with the WWW and perform the data input/output over the Web. For example, Lufthansa uses this to conduct auctions of last-minute flights (also refer to the "The Use of Information and Communications Technology in the Agreement Phase" subsection in Section 2.2.3). It is also possible to use interactive software. An example is the input acquisition and the shipping of standardized electronic business documents over the Web (also refer to Section 3.6). It is also possible to use the Net to access standard software such as SAP R/3 (refer to Chapter 4).

The information technical interconnection of the organizational units produces, in particular, the removal of previous media incompatibilities, an acceleration of information processes and a cost reduction for the data processing. It also helps increase the information transparency between the associated organizational units.

This background makes it evident that the Internet, as an important component of the information-logistical infrastructure, has a significant effect on the handling of business processes and thus on the form of modern coordination forms for independent organizational units. For example, the many subcontractors who are connected using the Net.

Intranet and Extranet

Intranet designates the company-internal use of Internet technology. The low-cost availability of the technology and the integration with Internet information offerings makes the creation of intranets attractive. Typical intranet applications are internal manuals, circulars, address lists, organizational guidelines and non-public parts catalogs.

In contrast, an Extranet is a closed network that uses the Internet to connect companies or a group of companies (for example, the connection of a car manufacturer with its suppliers or a producer with its logistics companies). As with intranets, extranets are based on Internet technology.

Computer Supported Cooperative Work

Computer Supported Cooperative Work (CSCW) is a research area within business data processing in which an interdisciplinary investigation is made how individuals cooperate in work groups, and how information and communications technology can be used to support them. CSCW includes the subareas of Workflow Management and Workgroup Computing. The following section briefly discusses these topics.

Workflow Management is concerned with the asynchronous coordination of cooperative working. The objects are work processes, such as order processing or an application for vacation, that are characterized by the forwarding of information within an intra-company or inter-organizational chain of processing stations. So-called Workflow Management Systems can be used to control, monitor and coordinate the complete workflow. An important task is the systematic handling of standardized documents. The workflow uses the network infrastructure to transport the documents. The central question is the standardization of documents that has been pursued since more than ten years under the term Electronic Data Interchange (EDI). The EDI standard, EDIFACT (Electronic Data Interchange for Administration, Commerce and Transport), describes a cross-industry and international standard for the data formats used to exchange business documents between companies. These documents are orders, delivery notes, invoices, customs declarations, etc. The goal is the automatic computer-to-computer exchange of such documents when the appropriate business situations occur (refer to Section 3.6 for a discussion of EDI).

Workgroup Computing designates the time-synchronous (but not necessarily location-synchronous) network-based support of work groups, such as project teams. For example, videoconferencing can be used for common sessions without requiring all participants to be present at a single location. The so-called Computer Aided Team Software provides a number of facilities to support group decision-making processes, such as the categorizing of the contributions that participants make on a black board.

Network Services

Internet providers offer private and company customers local access to the Internet. Private customers must usually pay a period-related fixed price and a fee that corresponds to the usage. Company customers often make a volume-related (for example on the basis of gigabytes) or capacity-dependent payment. Depending on the intensity of the usage, a leased line can be appropriate.

Larger companies and public institutions (for example universities), sometimes in the form of the appropriate subsidiaries, can also act as their own providers of access points. An Internet provider itself normally buys capacity directly from the network operating companies, such as Telekom and Arcor, or uses offerings from institutions, such as the Deutsche Forschungsnetz e.V (German Research Network). A trend in the telecommunication to reduce the contract durations and the brokerage of capacities can be observed (also refer to the "The Use of Information and Information Technology in the Agreement Phase" subsection in Section 2.2.3).

Information Services

Information services provide information services of a guaranteed quality. For example, they collect data during the so-called "tracking" of transmissions and make these documents available to their customers. The acquisition of the data can be followed, for example, by the automatic reading of a bar coding that is placed over a moving part of a conveyor belt. Information services also offer data processing services. For example, market research institutions, such as ACNielsen, evaluate the sales data from supermarkets to obtain the information needed to restock shelves (http://www.acnielsen.com). Another information service is the output of data over the Net to the customers. For example, standard software modules can be used to supply non-specific information to a customer inquiry, such as the display of the current item information for packages over the WWW ("tracing").

The quality of information services includes such things as

- a continuous, error-free operation of the server and the line
- short response times to customer inquiries
- user ergonomy
- short answer times and availability of qualified personnel for system problems

- state-of-the-art production of the promised and required information.

2.2.3   Electronic Commerce in the Supply Chain

"Electronic Commerce ... is the sharing of business information, maintaining business relationship, and conducting business transactions by means of telecommunications networks" (Zwass 1996). Because of the increasing general availability of low-cost Internet access and the sinking cost of the data transmission, the Internet, in particular, adopts an important role in the support of business relationships. We differentiate between three important areas of electronic commerce in the phases of a business relationship:

- the initiation phase (for example, supported by electronic product catalogs or electronic tenders)
- the negotiation phase (for example, supported by network-based auction procedures)
- the execution phase (for example, supported by network-based payments systems or through the direct (information) goods shipping using the Net).

The Use of Information and Communications Technology in the Initiation Phase

In the initiation phase, the requesting party identifies the requirement and searches for a provider. The provider will use the appropriate marketing to inform the consumers about its products and to find interested consumers.

Electronic product catalogs are an example of the information and communications support during the initiation phase. Such catalogs are multimedial interactive hypertext items that present and describe the products. The Internet contains many examples of electronic product catalogs. For example, potential customers can define the specifications for their next car on the Web-pages from Volkswagen (http://www2.vw-online.de/international/deutsch/products/frames_.htm) or BMW (http://www.bmw.de/produkte/index.htm), search the product database of the HotHotHot (http://www.hothothot.com) sauce manufacturer according to various criteria, or request product information and price for the wine offering of the Virtual Vineyards (http://www.virtualvin.com).

In addition to electronic product catalogs, inquirers can, for example, use search engines in the Internet to find potential business partners. These are large databases with special query technologies, such as Lycos (refer to http://www.lycos.de) and Yahoo (refer to http://www.yahoo.de). When search engines are used, they find only those participants that have a registered Internet presence.

Finally, biddings in the Internet, for example, can also support the initiation phase of a business relationship. An example here is Lawrence Livermore National Laboratory (LLNL), a producer of high-tech defense goods in California (Buxmann/Gebauer 1999): LLNL places contracts for the production of complex preliminary products on a (password protected) Web-page in the Net. Suppliers whose quality LLNL has previously approved are given an appropriate access right. These producers can make bids on this Web-page (they cannot see the tenders from the competitors). LLNL also places the product description as file with Computer Aided Design (CAD) data in the appropriate page. The producers can download this file and further process it in their systems. This procedure helps to accelerate the processes and to avoid misunderstandings between the project sponsor and the contractor. Producers that do not have the necessary Web access and the CAD programs will miss out on future work.

The Use of Information and Communications Technology in the Negotiation Phase

Discussions on the purchase and purchasing conditions take place during the negotiation phase of a business relationship. This means agreeing on price, quality, delivery dates, quantities, etc.

Electronic auctions are an example for a support of this phase using information and communications technology. The following example of a network-based auction system in the Dutch flower industry illustrates the principle of the method. The system based on a closed network using ISDN makes it possible to use a PC home to trade in flowers. The traditional Dutch Auction then takes place in which prospective buyers logon to an auction and the auctioneer specifies the origin, quality and minimum sales quantity for the flower quota. The auctioneer also starts a so-called auction clock running backwards, namely, his price suggestion, which was initially set high, decreases during the course of time. The first inquirer that reports over the network and thus stops the clock purchases a subquantity or the complete quota at the price shown by the clock when it was stopped. An associated logistics system handles the transport of the flowers from the seller to the buyer.

The success of this network-based auction system results from the auctioneer making a stringent check of the quality details of the flowers and also the strict separation of the price-setting process from the logistics processes that permits a significant reduction of the delivery times. The flowers to be auctioned no longer need to be brought to a centrally located auction hall, where the traders present at the auction can inspect the goods and submit their bids, but instead the flowers are transported directly from the producer to the buyer.

A prerequisite for this network-based auction is the trust the involved parties have that the information system actually shows the participant in the competition who stopped the auction clock first. In contrast, other auction procedures are not based on a specific inquiry at a specific time, but on supply and demand over a time period; for example the Vickrey Auction that is frequently mentioned in literature in connection with decentralized transport planning.
Other examples of the auctioning of goods and services are the second-hand goods auctions ONSALE (http://www.onsale.com) and EBAY (http://www.ebay.com), and the flights auction held by airlines, for example Lufthansa (http://virtualairport.Lufthansa.com/deutsch/lh_auction/auction.htm).

The Use of Information and Communications Technology in the Execution Phase

Financial and goods transactions are performed in the execution phase. The network-based payment systems that have been used for quite some time in the banking system (for example, SWIFT (Society for Worldwide Interbank Financial Telecommunication)) with closed networks can be used to provide the information and communications technology support for financial transactions. However, the use of the generally open medium Internet exhibits higher security risks (refer to Section 2.2.4) and demands a new quality of payment systems.

The literature differentiates between payment systems based on credit cards and bank assets (refer to Figure 2.17):

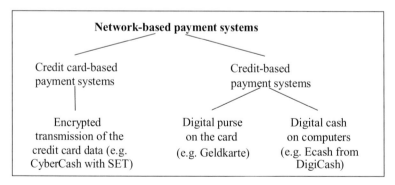

Figure 2.17: Types of network-based payment systems

In the following section, we first discuss the CyberCash procedure from the CyberCash Company as an example for credit card payment systems (refer to Figure 2.18). CyberCash works with the encrypted transmission of credit card data over all links and uses here the Secure Electronic Transactions (SET) transmission protocol developed by Visa and MasterCard together with well-known cooperation partners. (We discuss the basic principle of the encryption in Section 2.2.4.) The method is based on freely available software from CyberCash that controls the communication between the business partners.

Whereas, a customer in traditional businesses hands his credit card to the sales assistant, in an electronic purchase he sends his order and payment data encrypted by software together with the included electronic signature that identifies him to the trader (Step 1). The trader in turn forwards the received data including his signature to his bank (2) (Step 2 for traditional businesses corresponds to the transmission from the card reader). If the verification is positive (3), the customer's account is debited (4) and an appropriate message sent to the trader (5). When the trader receives the confirmation, he hands over the goods; the delivery can be made over the network for electronic products (6).

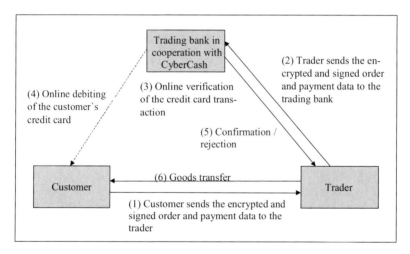

Figure 2.18: Encrypted transmission of data with SET as part of CyberCash (Stolpmann 1997, page 68)

Because in this procedure the financial institute that issues the credit cards to some extent is responsible for some of the losses that result from the incorrect use of credit cards, additional extensive registration processes for business partners with the concept from CyberCash can be avoided.

For payment systems based on bank assets, a differentiation is made between the use of a digital purse on the card and the use of digital cash on computers (also refer to Figure 2.19).

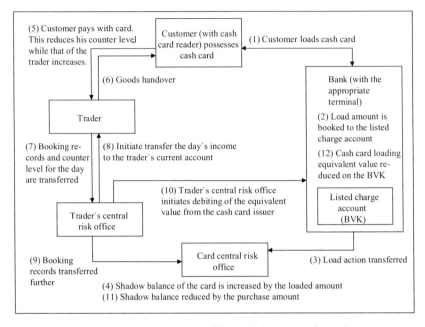

Figure 2.19: The concept of the digital purse on the card

A money card is, for example, an EC card provided with a computer chip that can be loaded at an appropriate bank terminal with an amount between 0.01 and 400 deutsche marks (1). The bank credits the loaded amount to an exchange clearing account (2) and transfers the load action to a card central risk office (3). The customer's balance on the shadow account for the card there is increased by the load amount (4). When a purchase is made over the network (5), the customer is requested to insert his money card in a card reader on his computer, which, for example, can be realized as slot in the diskette drive. Whereas the level of his card reduces here by the purchase amount, that of the trader increases. The trader then initiates the actual transfer of goods (6). The trader collects all the day's transactions and transfers the corresponding posting records to a trader central risk office (7). This initiates the transfer of the day's takings to the trader's current account (8), forwards the posting records to the card central risk office (9) and assures that the appropriate amount is collected from the money card issuer (10). Finally, the shadow balance of the card at the card central risk office is reduced by the purchase amount (11) and the money card equivalent amount on the exchange clearing account also reduced by this amount (12).

The concept of the digital purse on the card assumes a wide-scale acceptance of the money card and can operate with low transaction costs. This means that this procedure also permits a network-based payment of small money amounts.

The concept of digital cash on computers (for example, Ecash from DigiCash Company (refer to http://www.digicash.com) provides for the sending and processing of files that represent money amounts.

The goods stream is also initiated together with the finance stream. The support capability provided through the use of information and communications technologies can be differentiated into two cases. Provided only information goods are concerned, the complete shipping of goods can be handled over the network. For example, this permits the software industry to easily provide its customers with frequent updates or even complete software packages over the network. The situation is somewhat different for physical goods. Although the goods must be transported in the traditional manner, the networks can process the supporting services. For example, Federal Express has created a WWW service that customers can use to track their packages during the transport (tracking/tracing).

2.2.4   Security in Information and Communications Networks

The subject of security is of major importance in the public discussion of network-based systems. As in "normal life", there can be no absolute security in information and communications networks. If, for example, people with criminal intent wish to break in, irrespective of whether it is a building or a network, you can take preventive measures only to increase the effort they must expend to defeat the security mechanisms. Namely, when the time effort needed for a promising break-in attempt increases, the probability of detection before completion also increases. However, these measures consume resources and often reduce the user-friendliness for permitted users. The economic calculation states: minimize the sum of the costs for security system and the costs that result from the neglecting adequate security (for example, loss of assets by customers or damage to the image of a company that was broken into).

There are different security requirements and solutions in both closed networks and in the public network (also refer to Figure 2.20):

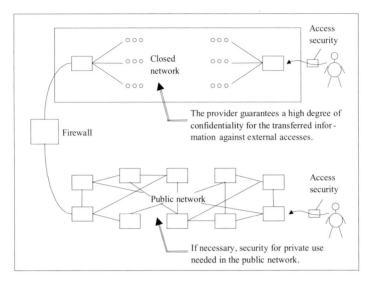

Figure 2.20: Security in closed networks and in the public network

Security in closed networks

Within a closed network, the network operator guarantees a high degree of confidentiality of the transferred information again attacks from outside. Security problems in such network can occur at the access to the individual terminals and in the transition into public network.

Generally, three different types of access security are used:

- check person-related criteria, such as fingerprints or face contours
- check owner-related criteria using hardware, such as with a chipcard
- check owner-related criteria using software, such as through the use of a password or a PIN (Personal Identification Number).

Although those access security mechanisms realized with hardware or check unchanging personal criteria exhibit the highest security standard, they are also the most expensive. They are used where there is a high danger potential or where there are large threatening losses, such as for the protection of central computers in large concerns. Although software-only solutions are cheaper, they have only limited reliability. In practice, mechanisms are frequently used that check the combinations of owner-related criteria using both hardware and software (for example check a card and a PIN). Furthermore, security officers use organizational regulations to increase the security, such as requiring that the password be changed every three months.

So-called firewalls are often implemented when there are interfaces between a closed network and the public network. These partition off an inner security zone from the public network within which potential intruders operate. This is done by monitoring the incoming and outgoing data packets, and rejecting unauthorized packages at the entry. Chapman/Zwicky (1995) provide further information about firewalls.

Security in the public network

The security problem in the public network results from the fact that many people use the largely unregulated medium Internet without being checked beforehand. From the user's viewpoint, the increased effort needed to ensure the security in the network must be related to the small cost for the connection and the low data transmission costs. The basic types of security problem occur in the public network:

- Targeted attempts of unauthorized persons to tap into or change a communication between participants. The following challenges must be mastered here:
- Confidential messages must be unreadable by third-parties (secrecy).
- It must be possible to determine that the consignor and consignee of a message are the only parties involved in the data transmission, and neither party has been misrepresented (authentication).
- A procedure must be found that ensures that the contents of the transferred data have not been manipulated during the transfer (integrity).
- Non-targeted attempts by participants to interrupt the communication in the network (for example, through sending viruses).

Cryptographic procedures are suitable to be used to ensure the secrecy and authentication. These procedures can be differentiated into symmetric and asymmetric encoding procedures. We describe first the simple *symmetric* encoding procedure. In this case, an original message M is encoded using the encoding procedure K. The recipient decodes the encoded message C using an inverse decoding procedure. A third-party, who does not know the encoding procedure agreed between the sender and the recipient cannot interpret the transferred message. A simple example: a text message is to be encoded with the procedure K = {use the next letter in the alphabet for each letter!}. Thus M = "Business" becomes the encoded message C = "Cvtjoftt". The associated inverse decoding procedure is {use the previous letter in the alphabet for each letter!}. The use of a symmetric key has the disadvantage that the involved parties must agree the encoding procedure beforehand using a secure channel.

The kernel of the *asymmetric* procedure, which eliminates this disadvantage, is the creation of two different, but closely related keys, for every participant in the network. One of the keys is made known publicly (for example on the homepage, as footer in the outgoing mail or even in "yellow pages"); consequently, it is known as the public key. The other key remains secret with the participant (for example stored on his smart card); it is known accordingly as the private key. To solve the secrecy task, the sender now uses the public key of the recipient to encode the document. The resulting encoded message C can only be decoded using the private key known only by him (refer to Figure 2.21).

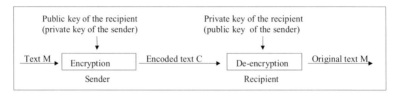

Figure 2.21: Asymmetric encoding (public key procedure)

However, the asymmetric encoding also has a disadvantage: it requires much more computer time than the symmetric procedure. Because of these problems, the two procedures are frequently combined, for example, this is case in the Secure Socket Layer (SSL) procedure. The symmetric keys are first generated at the start of a session and then exchanged between the participants using a secure asymmetric transmission. The message is then transferred using symmetric cryptography.

The authentication is realized using a digital signature. The sender encodes the document with his private key so that the recipient can decode the encoded message using the public key of the sender. The aim is to give the digital signature the same status as the hand-written signature.

Critical for the solution of the secrecy and authentication task using the public key procedure is to ensure that it really belongs to the specified sender. This assignment is confirmed through the specification of a certificate that contains the public key and personal data about the identity of the sender. A so-called trusted third-party issues not only the certificates but also handles the generation and administration of both keys. The certificates, for example, are made available by being stored in a public database; in contrast, the private key for the each digital signature is stored, for example, on a money card, and then deleted from all data held at the certifying site.

As part of the increasing privatization of traditional public service centers, companies should issue certificates for a fee (Rebel, etc. 1997). Some countries have laws to regulate the legal situation here. The Information and Communications Service Law (IuKDG) applies in Germany and states that the certifying centers themselves must by certified and supervised by a public authority, the so-called Root Certification Authority. This produces a certification hierarchy (refer to Figure 2.22).

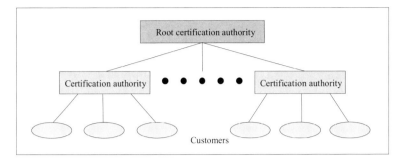

Figure 2.22: Certification hierarchy

The Root Certification Authority specifies the procedural rules that the certifying authorities must permanently observe. The Root Certification Authority is not allowed to issue certificates to customers and certification authorities are not allowed to certify other certification authorities. The Internet contains various national certification authorities such as the Trust Center of the Telekom in Germany   (Telesec;   http://www.telesec.de),   EuroSign   in   England (http://www.eurosign.com), and Verisign in USA (http://www. verisign.com).

Hash functions can be used to evaluate the integrity of a message. Such procedures are used primarily for the transmission of numerical data, for example account numbers. The sender uses hash function to create a check digit for a given document and sends this together with the document to the recipient. If, for example, the hash function used to generate a check digit for an account number is {form the sum of all digits and take the lowest-value position as check digit!}, an account number 123 456 789 receives 4 as check digit. The recipient uses the same procedure on the message and compares the calculated check digit with the received check digit. Because of the reducing nature of the formation of a check digit for a document, it is possible that different documents receive the same hash value. Thus, an unauthorized person could change a document in such a way that the recipient generates the same check digit as the original. However, the increase in the representation width for such check digits and the improved quality of the hash procedure means that the probability is very low, and in conjunction with the asymmetric encoding procedure (refer to next paragraph) can be further reduced.

The three previously mentioned procedures can, for example, be combined as follows: the sender first creates the hash code, uses its private key to encode the document together with the hash code (to authenticate itself), and then codes this message with the public key of the receiver in order to prevent other "eavesdroppers" on the Net from being able to interpret the transferred data. The recipient proceeds in the other direction.

Non-specific attempts by participants to interrupt the communication in the network concentrate on the creation and sending of viruses. Virus protection programs are used here as defense. These can often detect the virus attack by searching the delivered documents for all currently known viruses and, if detected, warn the user and offer the appropriate countermeasures.

2.2.5   Standardization and Network Effects

As in the area of the materials logistics, standards also adopt an important role for the information logistics. These standards normally use uniform rules to establish compatibility between system components. Thus, the materials logistics generally follow the goal of reducing the transshipment costs, whereas the information logistics aim at helping to reduce the information and communications costs.
In many cases, several standards that have a hierarchical relationship with each other are used together. For example, network protocols like TCP/IP define at the lower levels general rules for the exchange of information.

EDI standards for the transmission of commercial documents that build on these play an important role for the efficiency of logistics chains (refer to Section 3.6). A special feature of the assessment of standards results from these being so-called network effect goods. This means that the benefits of a standard for one user (for example, a cost-effective overall reachability of cooperation partners with a standardized system for the sending and receiving of documents) increases with the number of additional users. This provides a strategic problem for the software provider, namely, whether it should develop its own proprietary standards. In this case, a so-called de facto standard results through the momentum of a company in the market, for example the Business Application Programming Interfaces (BAPIs) from SAP (refer to Section 4.1). As alternative or in addition, providers use of open standards to make their systems compatible for external systems, for example XML (refer to Section 4.3). Previously, it could be observed that large providers, such as IBM, frequently attempted to force through their own proprietary standards, because in this way, in contrast to their competitors, they could provide their customers with network effects that offered a competitive advantage. However, times appear to have changed, when the current activities of IBM in the Java area and the availability of open interfaces in SAP systems are considered. This made possible the openness of cooperation between companies, such as those in the supply chain.

2.3 Supply Chain Management – Specification and Monitoring of Logistics Networks

Using the discussion of the materials logistics and information logistics as basis, we now consider the supply chain management as strategic management concept for the planning, control, realization and monitoring of the goods and information flows in a logistics network or to accompany the complete supply chain. Section 2.3.1 first describes the principles of supply chain management before discussing the concepts and instruments in Section 2.3.2. Sections 2.3.3 and 2.3.4 then provide a critical consideration of supply chain management, which first discusses the problems of the control of supply chains and then the problems with the cooperation of companies within several supply chains.

2.3.1   Principles of Supply Chain Management

Supply chain management designates the integrated optimization of the supply chain for the provision of processing material by various participants through to the delivery at the end-customer, possibly over several intermediate stages, of the manufactured product and associated services. This is based on the concept that the orientation on the needs of the end-customers to improve the service and to realize an uninterrupted goods flow and thus realize the fastest possible transport. In particular, the following effects are achieved:

- The knowledge of the complete value chain relationships as part of supply chain management permits the uniform planning and control of purchasing, production, storage and shipping for all companies. The shared operation of a single warehouse in the goods flow from the preproducer through to the final assembly company can, for example, reduce inventory levels and thus reduce the costs associated with tied-up capital.
- As result of the reduction of the throughput times (for example, through the use of modern optimization methods), the end-customer can be provided earlier with the product. There is also typically the aim to improve the ability and reliability of delivery and the delivery flexibility.
- There is also an improvement in the product and service quality that provides the end-customer with additional benefits, for example through the integration of necessary services in the supply chain, such as providing the customer with information for the consignment tracking.

The specification of an integrated value chain as part of the supply chain management enables products to be created with improved service and at lower costs. The integration here is made through the cooperation of all companies involved in the supply chain. The individual companies no longer compete on the market, but rather one supply chain (and the companies working together there) competes against other value chains.

2.3.2   Concepts and Instruments of Supply Chain Management

As previously discussed, supply chain management is based on the integrated planning, control, realization and monitoring of the information and goods flows along the complete supply chain, from the first supplier through to the end-customer. The integrating form of all process-coupled value activities while observing the effect of interactions between the transport and production activities require planning that goes over company boundaries. This requires a cooperation of all companies involved in the supply chain. The processes to be integrated have both a goods and an information dimension.

Note in the integration of the processes as part of the goods dimension that a cost reduction must always be assessed in conjunction with a total cost reduction within the compete supply chain. For example, this means considerations must be made whether the measures involving an inventory reduction in the own company (procurement warehouse) cause an increase in the inventory in the sales warehouse of the suppliers. If this case occurs, all supplies will match their prices over the long-term, which means that no positive effects can be expected from this restructuring.

In accordance with the concept of total cost, measures should be aimed at improving the logistics service in the complete supply chain and so produce an additional benefit for the (end-)customer. For example, a reduction of the transport time for a commodity from a suppler to a producer does not necessary produce a reduction in the total delivery time and thus an improved delivery service when this process is not coordinated with the production process of the producers and so results in wait times for the commodity there.

As Section 2.2 shows, the prerequisite for the binding specification of the goods flows over various companies is an integrated information processing that permits the best-possible preparation of the decisions to be made for the complete process chain. In the ideal situation, all internally and externally involved parties are connected with each other in real-time, and exchange needed information without delay. The automation of the data flows and the exchange of standardized formatted data, for example using EDI, makes the classic postal route increasingly superfluous. The computer-computer coupling from the consumer to the supplier, permits not only the automated data transmission, but also a mutual immediate access to the specific scheduling files. If the supply chain becomes interrupted, for example, as the result of the failure of a machine at the supplier, the system forwards the appropriate information to all parties involved in the supply chain and so permits all involved companies to initiate an alternative planning and so avoid the consequences of the failure, for example, by adding a new supplier. Currently, the discussion of the term "Cyber Logistics" concerns the use of Internet and Intranet solutions to provide new capabilities for the improvement and integration in the logistics chain.

Quick response systems are one example of the integrated consideration of goods and information flow. Scanner cash registers in retail outlets collect the sales data for various articles at the "point of sales" and forward these without delay to the upstream members in the logistics chain. The resulting improved sales forecasts mean, for example, that the deliveries from warehouses and branches to the wholesaler and the producers can be better planned. The integrated control of the complete supply chain permits a reduction in the inventory levels and in the throughput times, which in turn reduce costs, improves the delivery service through an improved availability of the products, and increases the quality of the products, through, for example, an improved product freshness in the consumer goods industry.

The Collaborative Planning, Forecasting and Replenishment (CPFR) concept is closely related to the quick response systems (also refer to Knolmayer, etc. 2000). This is a business model for the companies in a supply chain that provides a complete concept for supply chain management. It focuses on the "CPFR Voluntary Guidelines" that were prepared by the CPFR Committee (refer to http://www.cpfr.org). This committee consists of representatives from approximately 70 industrial and trading companies.

The CPFR aims at improving the inter-organizational partnership between suppliers and customers in the supply chain through the commonly-managed information and cooperatively-managed processes to result in a multi-winner game for the participants. These guidelines focus on reducing the inventory levels while improving the delivery service.

The most important guidelines of the CPFR concept are:

- The partners develop rules for the cooperation. These include agreements on the shared use of information and arrangements on the rights and obligations of the partners, and also for the criteria and metrics to be used to measure the effectiveness and success of the cooperation. In particular, a general plan is developed in which the participants are assigned main activities ("Core Process Activities"). The partners also develop a common business strategy, which results in assignments, for example, with regard to minimum order quantities and ordering intervals, and agreements on marketing and sales actions.

- The cooperating companies prepare together a forecast on the requirements of the end-customers. This common forecast can, for example, be prepared by combining the individual forecasts of the individual participants in the supply chain. The forecasts, can be determined, for example, using point of sales data from traders, goods that leave distribution centers, and orders arriving at the manufacturer. The shared forecast serves as basis for the plans and schedules to be agreed. Knolmayer, etc. (2000) provides an example.
- The partners attempt together to avoid unsuitable schedules in the goods flow. For example, if a trading operation rejects short delivery times and small order quantities, this can result in the producer in changing from manufacturing for stock to manufacturing for the customer, and thus can reduce the tied-up capital and the uncertainties associated with the scheduling of a warehouse for completed goods.

The following section discusses two examples of important instruments of the supply chain management, namely the use of standards in the information and materials logistics, and the use of modern simulations and optimization methods.

The implementation of quick response systems requires an unbroken, secure and fast information flow from the trader to the manufacturer to ensure the continuous supply and collection of sales data. The use of an EAN coding for the complete identification of the products and their packaging, and the use of uniform standards, such as the internationally used EANCOM format in the consumer goods industry, are two examples that support the specification of such an information flow. The use of standardized transport contains and loading equipment with the aim of providing a continuous and fast transport of the goods from the product manufacturer, over trading companies, through to the end-customer, are examples in the materials logistics.

The control and monitoring of logistics networks demands a permanent matching of the customer demands and produced-goods inventories. Should a bottleneck arise in the delivery to an end-customer, alternative procurement measures (for example, alternative transport facilities, other warehouses, additional purchases, etc.) must be checked (refer to Figure 2.23).

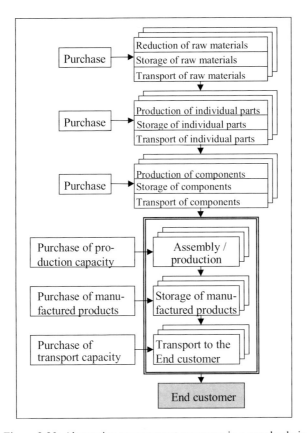

Figure 2.23: Alternative procurement measures in a supply chain

Delivery bottlenecks can have various causes. It must first be determined whether a capacity problem occurs in the transport of the completed products or whether insufficient inventory is available. In the first case, the purchase of additional transport resources should be considered. If the inventory is too low, check whether the missing goods can be obtained from other warehouses, whether you can produce the goods yourself or whether they should be purchased. When production bottlenecks arise through lack of capacity, a decision must be made between production at alternative production plants and the purchase of the missing production capacity. When production bottlenecks result because of missing components, it must be clarified for the components whether a capacity problem arises with the transport or whether insufficient inventory is available in the warehouse. This procedure requires that the planning of the procurement measures considers all the transfer and production steps of the compete supply chain from the mining of raw materials through to the transport of the completed products to the end-customer. The selection of a suitable procurement alternative is then made on the basis of cost criteria or the delivery service, etc., for example the delivery time.

Similar simulation and optimization methods can be considered as being the extended arm of the transactions systems, which permit a reporting, interactive and graphically presented decision support, and optimization, and indeed over the boundaries of a company. These procedures include items such as the preparation of demand and sales forecasts, and the planning of the production, the goods distribution in the network, locations and transport. The various planning modules act together and can simultaneously analyze the effects on other companies and divisions when a planning change occurs in a company or a division. The Advanced Planner and Optimizer is an example of such a simulation and optimization instrument. This tool is generally discussed in Chapter 6. Chapter 8 describes this in conjunction with the Goodyear case study.

2.3.3   Problems with the Control of Supply Chains

Supply chains comprising of various cooperating companies appear as consortium before the customer. The differences in the strategies of the value chain on the one hand and the companies integrated here on the other hand cause control problems for a supply chain.

The following questions give an impression of the task involved:

- How are the business risks assigned in a supply chain?
- How are the additional marginal returns assigned in a supply chain?

This raises the question whether it is possible to replace a producer's distribution warehouse and a procurement warehouse for a subsequent assembly plant with a warehouse operated together, and so reduce inventory held along the supply chain. Such changes frequently also have effects on both upstream and downstream companies in the supply chain. Which participants must now carry the inventory risks? And assuming that this measure produces a cost reduction and the resulting sales produce an increased profit: how will the resulting added-value be distributed?

There are currently no theoretically based procedures that have been proved in practice to answer these questions. Cases can be observed in business in which the most powerful company in the supply chain or, for example, the best-known brand to a customer, apparently "suggests with emphasis a distribution key" its special role to the other participants, which then permits these to have a more or less amenable working. The strength of the leading company can be used for such things as the disciplining of companies in the supply chain should these not observe the previously agreed rules. The probability of a satisfactory solution for everyone with this procedure is greater the higher the additional marginal return achieved with the (new) organization of the supply chain. It also increases in the case of this control model with the relative strength or visibility advantage of the leading company in a supply chain compared with the other participants.

Two examples for automobile manufacturing provide a general explanation of the relationships. In the "car delivery" supply chain in Germany, the manufacturer of a vehicle is frequently both the most powerful company and that which has branding advantage with the customer. Consequently, it is often the focal point of a automobile supply chain. It determine the strategy of the value chain, provides the implementation, monitors the regulated operation and organizes the adaptation of the value chain to meet the changing environmental conditions.

Newspapers have often reported on the procurement-side rationalization concept of VW under the executive López in the mid-1990s. Among other things, the automobile manufacturer provided the component suppliers with software for supply chain management. This software used the latest technologies to automate both the inventory management and the scheduling, and also gave the assembly plants a better view of the date and cost structure of the suppliers. Mercedes-Benz Also assumed the leading role in the procurement-side supply chain for the production of the Smart. The Smart production is based on new partnerships with suppliers and component suppliers, and on logistics that can respond flexibly to changes. Large module units are produced directly at the factory in Hambach in France by suppliers of systems and modules established by Mercedes-Benz. This avoids investments in conveyor systems, storage processes and the repacking of resources, and not only reduces costs but also avoids garbage. The close association of the component suppliers reduces the manufacturing depth of the DaimlerChrysler subsidiary Micro Compact Car to ten percent. This greatly reduces the required investments and limits the risk for the DaimlerChrysler subsidiary.

A possible alternative concept of the management of a supply chain is through the specific use of logistics and service competence, namely central skills in the value chain. In particular, well-organized alliances of the mid-sized companies have chances to "emulate" the capability of large concerns with a network of participants without making the usual head office charges for the financing of the coordination and without needing to pass the correspondingly higher prices on to the customer (refer to Section 2.4).

2.3.4   Problems with the "Coopetition" of Companies

Companies often pursue the goal of avoiding a one-sided dependency on a monopolist: Such a dependency can be a component supplier or a customer. This means that many companies act as participants in various supply chains. This in turn means that two companies that cooperate as part of a supply chain, can also be competitors in two other value chains. English language literature has coined the term "coopetition" (from the words "cooperation" and "competition") for this situation. How can a company master this challenge?

There are also here currently no theoretically-based procedures that have been tested in practice to answer these questions. In principal, it concerns the interconnection of various supply chains: What positive network effects can be ideally achieved for all participants through the connection of the processes in different value chains? This question arises because companies are increasingly concentrating on a few central competencies, thus the number of companies in a supply chain grows (for example, "delivery of a vehicle"). Cases can be seen in business, in which, for example, a manufacturer builds the motors from a competitor into its cars. Apparently a differentiation is made in various supply chains between the management roles and "cooperative" roles decoupled using clear interfaces. These prerequisites permit, for example, a motor manufacturer to increase the utilization of a new factory by supplying competitors to such an extent and achieve an improvement of the return on investment so that any possible growth limitations are overcompensated in its own end-product market.

2.4 Modern Tasks for Logistics Service Providers in the Supply Chain

This section describes modern tasks of the logistics service companies that play a special role in the supply chain. Section 2.4.1 first emphasizes the necessity for logistics service providers to offer system services. Section 2.4.2 then describes the capabilities of the service management for logistics service providers, before describing in more detail in Section 2.4.3 the role of the logistics service provider as supply chain manager.

2.4.1   From Logistics Service to Logistics System Service

Because logistics service providers have central competencies in the logistics processes and provide good prerequisites for a specific expansion of these in accordance with modern demands, they play a special role in the supply chain management. For the organization of consolidated production systems with worldwide manufacturing locations, the logistics cannot be restricted to just organizing transport and managing inventory. Rather, what is needed is an interconnection of all logistics processes and their resulting processes to form a transparent and thus controllable, thin value chain. The task is to further develop to become the system service provider with strong orientation on customer needs and with excellent service provide comprehensive business products that satisfy the various customer requirements.

With this knowledge in mind, logistics service providers, in particular, increasingly offer system services, which, in addition to the logistics central services of transport, storage and transshipment, can contain a range of further services (refer to Figure 2.24).

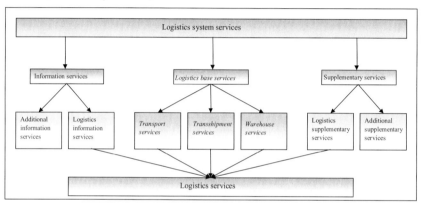

Figure 2.24: Task areas for logistics service providers (Isermann 1998a, page 15)

These logistics system services consist of so-called supplementary services that can be logistical and additional. Logistical supplementary services are present when an additional service permits or makes more economic the central services shown in Figure 2.24. These include, for example, packing or picking. Thus, these are supporting processes in the goods flow. In contrast, any supplementary services provided in conjunction with a logistics central service are just an additional supplementary service. An example is when a clothing shipper undertakes not only the transport and storage, but also the ironing of the outer clothing, or a logistics service provider that assembles the collected parts from the suppliers to make a system part.

Logistics service providers also offer so-called information services that are subdivided into logistical and additional. Logistical information products designate those services that directly control the goods flow. They are used to collect all the information needed to create, and process logistics service, and to make it available where it is needed (for example, the sending of transport information to a distribution warehouse using EDI). If, however, the information generated for the logistics service is made available to the customer, this is an additional information service. Examples are the analysis of the ordering behavior of the customer using an ABC analysis (refer to Section 2.1.3) or as the provision of information for consignment tracking for the customer using the Internet (tracing).

2.4.2   Service Management as Modern Task Area for Logistics Service Providers

The cost and service goals of the logistics services should be assessed to the extent to which they meet the customer's needs (for example, the organization of the transport). Operating numbers can be used to help make the assessment. Such operating numbers are easily collectable indicators in aggregated, quantitative form, which, for example, represent the current state of the service for departments and thus provide the capability to rate comparable organizational units. The logistics service provider in the service management can orient on the delivery service as indicator for the efficiency of the service products. The delivery service consists of the delivery time, delivery reliability, delivery satisfaction and delivery flexibility components (refer to Figure 2.25).

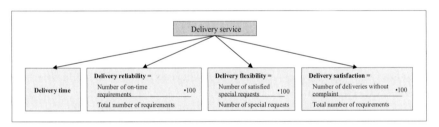

Figure 2.25: Service-related operating numbers

- The delivery time designates the time range between when the customer places the order and when it receives the goods. In particular, it consists of the times required for the order processing, production (provided no delivery can be made from inventory stocks), picking, packaging, loading, and transport.
- The delivery reliability is the ratio of requirements made on time to the total number of requirements. A requirement can be defined as either an order item or complete order, and, in particular, as quantity or value. An increase in the delivery reliability can be achieved, in particular, by increasing the inventory levels for a part.
  To determine an optimum degree of delivery reliability, the costs defined as the operating figures that arise when there are missing items (for example, the opportunity costs resulting from lost orders) are compared with the addition warehousing costs resulting from an increased safety stock level. When the

delivery reliability increases, the costs resulting from missing items tend to decrease and the warehousing costs increase. The optimum delivery reliability level then results from minimizing the total costs (which consist of the sum of the costs for missing items and the storage costs).

- The ratio of the number of satisfied special wishes to the total number of special wishes serves as indicator for the delivery flexibility, namely the capability of the logistics system to respond to special requirements from the customers.
- The delivery satisfaction can be determined from the relation of the number of complaint-free deliveries to the total number of requirements. Complaints here are any objections concerning the type or quantity of the delivered goods, and damaged or defective goods, but does not concern the delivery time. This operating number also records all complaints resulting from deliveries to the wrong location.

How the delivery service combines the individual components and how the components are to be weighted depends on the individual wishes of the customers. For example, a shorter delivery time by itself does not increase the delivery service when the customer did not request this or an earlier delivery indeed is not desired.

However, the service management as modern task area for logistics service companies does not only cover the optimization of the central logistics services with regard to service aspects, but, in particular, also the range of additional services (additional supplementary services and additional information services), and thus logistics system services. Additional services are service products, which, for example, Schenker offers for the "Finished Vehicle Delivery". Vehicles are subjected to a quality check during the transport and before their delivery to the customer. This quality check takes the form of a visual control for any paint damage. At Hewlett Packard, for example, the shipping companies perform the non-technical installation work for the distribution of networks. Logistics service providers can, for example, also perform after-sales services, such as a spare parts supply or repair services during the guarantee time or support time, and provide billing services as additional supplementary services.

Industry and trade increasingly want to cooperate more with logistics service providers that are qualified to be used in complex logistics chains. The service spectrum for logistics service requires here not only the scope of a system service but also demands a worldwide orientation. The increasingly prevalent term "global player" is already satisfied by some companies. Global players have greatly extended the scope of their service products and include partners in such concepts when it is not possible to provide of logistics services with their own resources or own infrastructure. The continuously progressing development of the information and communications technology permits here the necessary horizontal cooperation to provide a worldwide supply to the customers as part of a "door-to-door service". Only a worldwide service association can satisfy the requirements of future global procurement, production and distribution tasks.

2.4.3   The Logistics Service Provider as Supply Chain Manager

The creation of supply chains increasingly demands a coordination that goes over company boundaries. Logistics service companies that do not consider themselves to be just "architects of the traffic" but as helmsmen of complete value chains can undertake such a task. Such logistics system providers plan, structure and control the complete logistics network in cooperation with service providers by making use of all supplier, branch and service provider locations where goods can be stored or transshipped, and so present themselves as supply chain manager. This implies the selection and inclusion of the companies to be integrated in the supply chain, the supervision of these organizations, and the use of disciplinary measures, such as the exclusion from the supply chain in an extreme situation.

The most important task for the managing service company is the structuring and control of an uninterrupted goods flow, and a customer-oriented service bundle using a powerful and complete information management. This initially includes the implementation of a inter-organizational information and communications infrastructure. This can require the

- the creation of powerful computer centers for the management of data and the execution of optimization tasks
- data connections between the companies linked in the network or the provision of connections to existing systems,
- the implementation of the appropriate standards as part of an interface management
- the provision of a server to collect all relevant information and to provide it, where possible, in real-time to the companies integrated in the logistics network
- to equipping of the companies attached to the network with software, hardware and, when necessary, also with personnel capacity for the data processing

All information and communications systems must be provided either by the logistics service provider itself or be obtained externally. If a suitable information infrastructure has already been implement for the logistics, the logistics service provider can use it to control supply chain. Thus, for example, tracking using scanner systems can be used to check schedules for service requirements and initiate the appropriate steps should there be discrepancies from the plan.

However, the supply chain manager uses the information infrastructure for the logistics not only as instrument to coordinate the supply chain but also provides a principal prerequisite for the cooperation of the companies involved in the supply chain and thus enable the supply chain management. The information infrastructure must permit complete information transparency in the supply chain. This permits all participants to supervise the actions of the supply chain manager and so provides trust in the managing company and the supply chain strategy it follows.

2.5 Supply Chain Management Systems

As discussed previously, information logistics provides the basis for an efficient planning, control and monitoring of goods flows in a logistics chain. Thus, the use of information and communications technology adopts a key role for the form of the supply chain. This becomes also apparent because many projects nowadays are implementing supply chain management systems in practice and standard software providers are also increasingly offering such systems. This section provides a short summary of the currently available supply chain management systems before the following chapters concentrate on SAP systems.

The systems normally cover at least the following planning tasks and functions, which, to some extent, build on the methods presented in this chapter:

- requirements and sales planning
- location planning
- vehicle routing
- available to promise
- production planning
- detailed planning.

Thus, one of the focal points of these systems is the support for the detailed planning functions both at the inner-company and the inter-organizational level. Both exact and heuristic procedures are used here. In addition to in-house developments, many software providers – including SAP – use standard libraries for algorithms, such as those from ILOG (www.ilog.com) and CPLEX (www.cplex.com). Another characteristic of many supply chain management systems is the use of storage-persistent data storage within main storage. This provides a speed advantage, which is of particular importance when used for optimization and simulation procedures. The best known providers of supply chain management systems are i2 Technologies (RHYTHM), Manugistics (Manugistics 6), Synquest (various tools), Logility (Logility Value Chain Solutions) and Numetrix (Numetrix/3).

Because this is a very lucrative market, the leading standard software manufacturers have recently started to provide such supply chain management systems. These systems have the major advantage that there exist clean interfaces to the corresponding standard software solutions. This is of importance because the planning functions for the supply chain management systems build on data that are normally available in the operative systems.

For example, Baan has developed SCS (Supply Chain Solutions) as solution. J. D. Edwards also offers a solution with the name SCOREx (Supply Chain Optimization and Real-Time Extended Executions). The Oracle Applications standard software has a Supply Chain Management module. Finally, Peoplesoft offers a solution with the Supply Chain Planning.

This book concentrates on SAP systems that support supply chain management. These are the systems

- SAP APO (Advanced Planner and Optimizer)
- SAP LES (Logistics Execution System)
- SAP Business-to-Business Procurement

which are part of the SAP Supply Chain Management Initiative. These systems use for their base data, for example, the data from the R/3 System or the data stored in the Business Information Warehouse (BW).

However, before we start with a detailed discussion of the use of the mentioned SAP supply chain management systems, Chapter 3 provides an overview about the mySAP.com initiative, Chapter 4 then initially describes the classic R/3 functionality for the support of logistics processes. Chapter 5 then discusses the possibilities of the use of the Internet as basis for inter-organizational processes and cooperation on the basis of SAP systems.

Chapter 3 mySAP.com

mySAP.com is an initiative and a marketing strategy from the SAP Corporation that not only links all previous system solutions with each other but also opens new business areas. The open software environment at the center of mySAP.com integrates intra-company and inter-organizational business processes on a single platform, the Internet.

In the following sections we describe

- the mySAP.com Marketplace that opens the use of an Internet portal
- the mySAP.com Workplace that provides a user with integrated access to various SAP and non-SAP applications (single sign-on)
- the offering of predefined mySAP.com solutions for standardized business scenarios that connect various SAP solutions with each other
- the possibility of Web-based Application Hosting that permits the cost-effective operation of various SAP systems
- the SAP technical infrastructure called Internet Business Framework, which is realized on the basis of the previously mentioned areas of mySAP.com.

mySAP.com represents a comprehensive concept whose parts SAP is implementing stepwise. Not all the concepts described in the following sections are yet available.

3.1   mySAP.com Marketplace

Since October 1999, mySAP.com Marketplace (www.mySAP.com) provides current and potential customers with an Internet portal. The long-term aim is the automated handling of inter-organizational business processes. In the current state of development, SAP is building the information base for the network market activity. This covers four central areas:

- Industry and product directories provide knowledge about contact capabilities and offerings. SAP calls these Internet Communities (refer to Figure 3.1).
- In addition, mySAP.com provides links to services, such as the subject areas of travel, languages and finance. Job openings can also be found here.
- Directories of important SAP business partners, i.e., companies that offer solutions in conjunction with SAP (e.g., software vendors, consultancies), and so-called marketplace partners, which, in cooperation with SAP, build and expand the marketplace offerings.
- You can also access SAP-specific services, e.g., product documentation.

Thus, the mySAP.com Marketplace designates a portal as a central entry and navigation point that connects various providers and inquirers with each other, and supports the users´ search through a range of distributed offerings. The users of portals are differentiated according to task or geographic criteria. Whereas task differentiations focus on professional or industry-related interests, the geographic forms concentrate on regional subject areas.

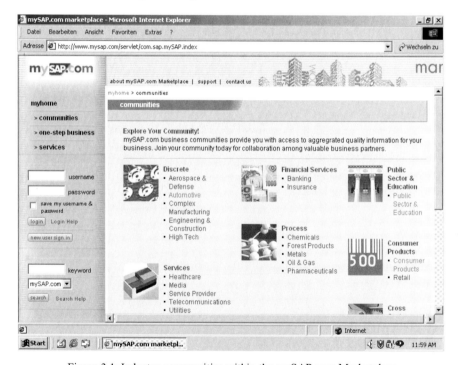

Figure 3.1: Industry communities within the mySAP.com Marketplace

The mySAP.com Marketplace allows an

- cross-industry horizontal
- vertical, and

- regional

differentiation of target groups.

Cross-industry marketplaces provide horizontal applications, such as auctions or business transaction. In addition to the trade in cross-industry goods and services, this provides general information such as messages, company profiles or exchange listings. In contrast, vertical marketplaces are based on industry-specific procedures and use, for example, uniform standards for the data transfer. Regional marketplaces can bring together providers and inquirers in geographically restricted areas.

For example:

Figure 3.2: Marketplace for the Chemical and Pharmaceutical Industy
(http://mysap.com/chempharm/en/index.htm)

The industry-specific marketplaces are starting in the area of the chemical and pharmaceutical industry (refer to Figure 3.2). Leading manufacturers, such as BASF, Bayer, Degussa-Hüls or Henkel, are currently working together with their suppliers on an open electronic marketplace for the procurement of materials, supplies and operating supplies on the basis of mySAP.com.

> *The central concept is an industry-wide standardization of procurement and sales processes. For this purpose, various business scenarios are supplemented with industry-specific information, which ranges from company-specific information through to messages and background information for the complete industry.*

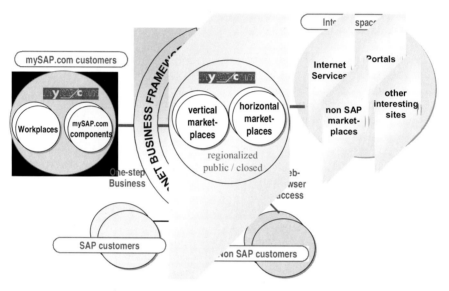

Figure 3.3: Structure of mySAP.com Marketplace

The portal provides companies which use standard SAP solutions (refer to Chapter 4 to 6) the capability to couple company-internal systems with the marketplace and thus extend business processes to the Internet. In contrast to conventional Internet-based solutions (refer to Chapter 5 - SAP Internet Applications), this should permit collaborative business relationships between different marketplace participants in the future.

Figure 3.3 illustrates the connection of different solutions to the mySAP.com Marketplace. SAP systems can use a technological infrastructure, called the Internet Business Framework (refer to Section 3.5), to connect to the marketplace. A coupling with non-SAP systems is being pursued.

The mySAP.com Marketplace uses Secure Socket Layer protocols (refer to Chapter 2) to transfer sensitive company data. A trust level concept is used to control the monitored access to the various areas within the marketplace. Information and services are classified and different authorization levels or prerequisites assigned. This permits information offerings to be viewed without the identification of the user. However, for example, a user registration may be needed to access financial services. The possession of a certificate is also necessary if a user wants to act as provider or inquirer of products. At the highest level, authorized market participants can access inner-company systems of an external partner. All marketplace transactions are also logged and saved with a special mySAP.com Message Storage Service.

3.2   mySAP.com Workplaces

SAP designates a mySAP.com Workplace as an enterprise portal that provides users with standardized access to inner-company and external SAP and non-SAP solutions. In particular for companies with heterogeneous system infrastructures, this provides advantages because different systems can be started from a single user interface taking account of authorization concepts.

A number of properties characterize the structure of the mySAP.com Workplace. In particular, the Workplace concept's use of the open Internet HTML standard as a front-end protocol satisfies the requirements of thin-client computing. The basic idea here is that application systems run on a remote server and the local thin-client uses the network, for example, to receive and display screen data calculated on the server or to send input data to the server. The user-side access on the Workplace is made with a XML-capable Web browser (refer to Chapter 5) and does not require the installation of a special client. The server-based administration of the solution and the use of standard browsers are also important elements of the Web-based Application Hosting (refer to Section 3.4).

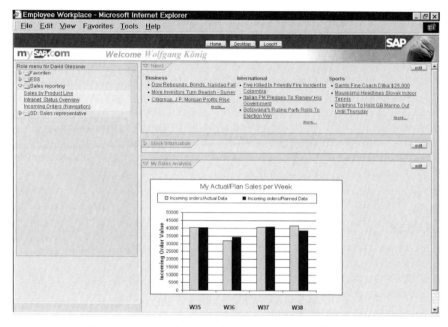

Figure 3.4: View of a preconfigured personalized Workplace

In accordance with the different roles of individual employees within a company (such as buyers, controllers or sales managers), the mySAP.com Workplace must be preconfigured and personalized. For example, the role menu of a sales manager can provide sales statistics and information about sales software or product catalogs. Furthermore, Figure 3.4 shows links to the business (e.g., stock exchange data), international (e.g., news) and sports areas.

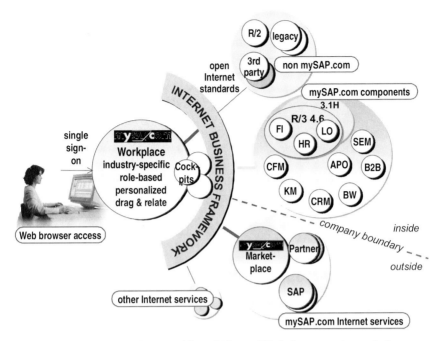

Figure 3.5: Access of a user with mySAP.com Workplace to various solutions

Figure 3.5 illustrates the access of a user with mySAP.com Workplace to various structured information items (e.g., R/3 data) and Internet resources. The Workplace also provides a complete integration with the mySAP.com Marketplace. The so-called Internet Business Framework (refer to Section 3.5) is used to integrate all the systems.

3.3 mySAP.com Components and Business Scenarios

SAP considers a business scenario to be the combination of various software modules and the resulting cross-system process form to provide a solution to a standardized business situation. For example, as part of the supply chain management, outside suppliers use the Vendor Managed Inventory to automatically monitor and refill the inventory of a company. Another example is the Supply Network Planning that optimizes the availability of individual goods in a distribution network (refer to Chapters 6 and 8).

Various SAP-specific and non-SAP components can be combined in mySAP.com business scenarios. In addition to the modular components of the R/3 System (refer to Chapter 4), the following components (refer to Chapters 5, 6 and 8) can be used:

- Business-to-Business Procurement and Selling (B2B)
- Business-to-Consumer Selling (B2C)
- Advanced Planner & Optimizer (APO)
- Customer Relationship Management (CRM)
- Business Information Warehouse (BW)
- Strategic Enterprise Management (SEM)
- Corporate Finance Management (CFM)
- Knowledge Management (KM)

Despite the extensions provided with mySAP.com for the Internet Business Framework and the inclusion of all solutions in this complete marketing strategy, the various software units are still distributed as independent solutions.

3.4   mySAP.com Application Hosting

The terms Application Hosting and Application Service Providing have not previously been finely differentiated. You can consider Application Hosting to be the operation of software licensed to a company in an external computing center. In contrast, Application Service Providing designates the offer of an external computing center to bill for services for software licensed to the service provider based on its use. In both cases the server hardware belongs to the external service provider. In particular for mid-sized businesses, this makes it possible to avoid extensive investments in system components and personnel.

mySAP.com Application Hosting provides a range of services, e.g.,

- Test Drive Your Solution Online: Test an Internet Demonstration and Evaluation System (IDES) before finalizing the contract.
- Compose Your Solution Online: Select functions and solutions in accordance with a company's requirements.
- We Implement Your Solution Online: Customers can use the Internet to implement solutions in the Service Computing Center.
- We Host Your Solution Online: For a fixed monthly fee, the Service Provider manages SAP solutions.
- We Build And Host Your Business Community / Marketplace Online: The service provider, in conjunction with SAP and customers, builds and operates marketplaces (refer to Section 3.1). This also applies for so as part of vertical marketplaces.

SAP has founded a subsidiary with the name mySAP.com to provide Application Hosting support.

3.5 mySAP.com Internet Business Framework

The exchange of messages as part mySAP.com is based on the Internet Business Framework (IBF) that uses various open Web standards. This is a infrastructure that uses the XML format description language (refer to Chapter 5) at the various layers of the cross-system communication (refer to Figure 3.6).

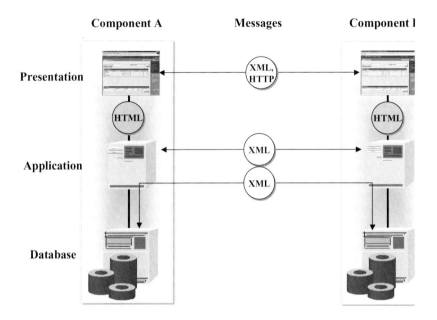

Figure 3.6: A Web standards-based infrastructure

This middleware solution is available for employees and SAP customers on the SAPNet.

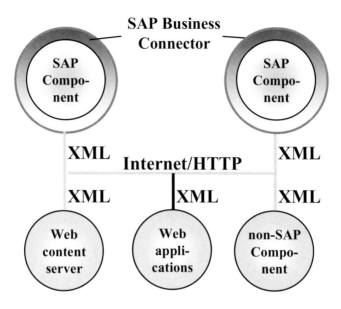

Figure 3.7: XML-based Web messages across the board

To simplify the message exchange between different system interfaces, SAP offers an XML-based Interface Repository. In the near future, this repository should provide a large number of interface definitions (e.g., BIZTALK, OAG, RosettaNet, W3C and XML.ORG) in order to simplify communication with external partners. SAP customers have the opportunity to create their own libraries for user-written or externally procured interfaces.

Chapter 4 Support of Inter-Organizational Logistics Processes with the SAP R/3 System

4.1 SAP R/3 in the Logistics: Overview

The R/3 System is a client-server solution for the support and automation of business processes in various industries. The software architecture is based on the separation of the database, application and presentation areas. The R/3 System provides processes and data structures for various functional areas, which can be customized to meet the specific requirements of the companies.

Together with accountancy, the support of the supply chain is a central application area for the R/3 System. SAP currently provides the following modules:
- Materials Management (MM)
- Plant Maintenance (PM)
- Product Data Management (PDM)
- Production Planning (PP)
- Project System (PS)
- Quality Management (QM)
- Sales and Distribution (SD)
- Service Management (SM)
- Logistics Information System (LIS).

The (standard) processes here, such as production planning (Knolmayer et al. 2000), use the functionality from various modules. A shared data base is used to integrate the processes, using a relational database that the applications can access. For the logistics area, the LIS consolidates the operative logistics data from various information systems present in the R/3 System and also provides the capability to evaluate the data. The LIS information systems provide a number of planning and forecasting functions, and the functions to manage the planning data. In addition to standard analyses, such as ABC analyses and planned-actual comparisons, it is also possible to produce customized reports. Predefined diagrams (e.g., barcharts) are provided to display the results, which can be further processed using graphics tools. LIS also contains an early-warning system that signals when certain threshold values are exceeded. SAP provides the Business Information Warehouse as another facility for the grouping, aggregation and analysis of company-internal and company-external data (refer to Chapter 6).

This chapter provides a more detailed investigation of those parts of the R/3 software used to support logistics processes that are closely related to processes or cooperation between companies. For example, the MM module covers major parts of the procurement logistics described in Chapter 2, whereas the SD module represents many components of the distribution logistics. As our case study in Chapter 6 shows, the MM and SD R/3 modules provide the basis of the R/3 System for logistical processing between companies. The service management already discussed in Chapter 2, supported by the SM module, plays an important role for processes with customers between companies. In contrast, the modules such as QM, PM and PP are primarily used internally in companies, even when they could provide particular benefits in the production planning through the inter-organizational cooperation. This chapter discusses only the MM (Section 4.2), SD (Section 4.3), and SM (Section 4.4) modules. The SD and SM modules also form a major basis for SAP's Customer Relationship Management Initiative, which focuses on the business processes with customers and is presented in Section 4.5. As previously mentioned in Chapter 2, EDI is one of the most important concepts to the support of inter-organizational processes in the supply chain. Section 4.6 discusses the capabilities and limitations of the realization of EDI on the basis of SAP R/3.

4.2   MM Module Functions

The MM module is used to support the complete procurement process, starting with the demand, through to the arrival of the goods in a company. It provides the following functionality

- Materials planning
- Purchasing
- Inventory management
- Warehouse management
- Invoice verification.

The following sections briefly describe this above mentioned functionality of the MM module. Keller/Teufel (1998) provides a detailed discussion.

Materials planning

The main task of the materials planning is the monitoring of inventory and, in particular, the generating of order requests for the purchasing department. The initiation of an order at the optimum ordering time, for example by comparing the inventory level with predefined parameters such as the reporting level and the minimum inventory level can the be made by a person or with an automatic procedure (refer to "Inventory and stockordering policies" in Section 2.1.3). In addition to such consumption-controlled material scheduling, which also determines the optimum ordering quantity, the R/3 System provides plan-controlled material scheduling.

The plan-controlled scheduling is frequently used to compare more expensive materials that either cannot be stored or stored only in small quantities (A-articles). It considers both the customer orders to be processed in the near future and the existing material reservations. It assumes that the primary requirement for end products and, for example, assemblies and individual parts that can be sold in the spare parts business is known and can be scheduled. Bills of materials are used to determinate the necessary quantities of the individual parts and assemblies to be used.

Purchasing

The support of the purchasing department covers the determination of potential supply sources and their assessment, the request for tenders for the order, the posting of the order confirmation, and the monitoring of goods deliveries and payments (Keller/Teufel 1998). All business activities of this functional area are represented using the document principle. This means that every activity is documented with a receipt. Each receipt consists of a header (identification number, general data such as details concerning the supplier) and several items (material, quantity). In addition to order requirements, inquiries, offers and orders, this principle is used to represent all goods movements within the R/3 System.

Inventory management

The business task of the inventory involves the acquisition of the value and quantity of the inventory stock. It includes the planning, acquisition and the recording of all goods movements, and also the stocktaking. The material stocks and their changes resulting from goods receipts and issues, returns and rearrangements are recorded in real-time and permit checking and correction to be made to the material flow. The inventory management also permits balance assessment and other common stocktaking methods, such as the stocktaking at a fixed date and the permanent stocktaking, and provides acquisition aids for their execution (Rebstock/Hildebrand 1999).

Warehouse management

The warehouse administration plays a key role in logistics. This includes the management of the warehouse structures, such as consignment warehouse, supply warehouse and distribution centers, and the warehouse types present there, for example, high-bay racking warehouse, fixed position warehouse and bulk storage. When needed, warehouse types are subdivided into individual warehouse areas, which in turn consist of a number of warehouse bin locations. The warehouse management can be adapted to the prevailing conditions, for example, to observe safety regulations for the handling of dangerous goods. The R/3 System also permits the common administration of geographically separated warehouse locations.

Invoice verification

The invoice validation has the task to check incoming invoices with regard to their content, calculations and price details, and so prepare for the payments. It contains functions to enter invoices and to represent the various types of tax.

4.3   SD Module Functions

The following section summarizes the functionalities of the SD module. The module SD supports the complete sales process, starting with the pre-sales support, including the order arrival, through to the completion of the delivery, and contains the function areas

- Sales
- Shipping
- Billing
- Sales support.

As with the MM module, Keller/Teufel (1998) provides a detailed description.

Sales

Within the module SD, the sales covers all activities from the acceptance of a customer's inquiry, including the preparation of an offer, through to the processing of an order (including the checking of creditworthiness and goods availability). Offers are supplemented with the necessary item data, such as materials, services and prices, and result in specific orders, which, depending on the customer relationship, may have different terms and conditions. General contracts cover long-term customer relationships and may be differentiated according to contractual conditions and delivery schedules. Whereas contractual conditions cover quantities and prices, delivery schedules contain delivery quantities and delivery dates. With particular regard to the specific requirements of industries, for example the automobile industry, the handling of the delivery schedule constitutes a significant part of a JiT supply to factories. The material specified in a delivery schedule is requested using assignments, which, in conjunction with EDI, results in an automatic update to the system.

As our case study in Chapter 7 shows, it is possible here to pass new requests to component supplier companies, which is then known as a JIT delivery schedule.

Shipping

The shipping performs the provision of all articles contained in an order. The delivery documents passed as reference copies from the order processing and processed with the aim of the dispatch optimization initiate the dispatch processing (refer to the case study in Chapter 8). The dispatch covers activities such as the planning and monitoring of the work effort, various internal processes such as picking and packing, the preparation of dispatch papers and monitoring of the delivery until its arrival at the customer.

Billing

The billing represents completion of a business activity within the SD module. It contains functions to support the processing of invoices, credit notes and debit notes, and pro forma invoices. The invoice can be created as an individual invoice or a collective invoice. For example, it is possible to combine different deliveries for various orders as a single collective invoice. It is also possible that different deliveries of a single order result in individual invoices. It is possible to create so-called invoice lists when an order consists of items from different customers.

Sales support

The sales support provides marketing and sales relevant information with interfaces to other companies. This includes information about suppliers, customers, partner, potential interested parties, competitors and their strategies, and products. It is closely associated with the areas of sales, shipping and invoicing, and has the aim of providing support to employees for the acquisition and support for suppliers and customers. The gathered information serves as basis for the preparation of tenders and orders, and is collected as part of the sales information system.

4.4 SM Module Functions

The Service Management provides several functions that cover various scenarios for the customer service. The complete function area contains the
- Technical objects,
- Warranty,
- Contracts and Planning,
- Create service notification,
- Service processing,
- Deliveries of spare parts,
- Billing, and
- Information system.

Technical objects

Prerequisite for an IT-supported service management is the representation of technical objects, for example photocopiers and vehicles, at customer sites. Because the structures of these customer installations differ greatly, service objects can be represented as either instances or types. In addition to technical information, such as serial numbers, assignments to organizational units or customers' employees can also be established for each object. This can be used, for example, to define a contact partner for a specific object.

Furthermore, complex interconnected structures that consist of technical locations or equipment, can be entered and represented together with their dependencies. They belong to the master and structure data of the R/3 System.

Warranty

Warranty require the service providers to offer a service to customers within a time period or for a specific usage duration without any total or partial payment. The SM module provides support for time-related warranties (for example, three months), counter-dependent guarantees (for example, number of driven kilometers) and warranties for time periods and time intervals (for example, 100,000 kilometers in a maximum of five years). These restrictions are checked automatically as the result of a customer message or for an invoice request.

Contracts and Planning

Service contracts can be used to specify long-term agreements with customers concerning the scope and the billing of service work. A general distinction is made here between maintenance contracts and leasing contracts. Such contracts describe which services are to be provided for which objects at which terms, and can be either short-term or long-term. In the case of periodic services, references are made between the contract and a maintenance plan to ensure an integration between contractual agreements and the processing. The invoicing department uses the service contract information to check whether and in which form the customer is to be billed for a service.

Create service notification

The SM module provides functions that can be used to enter and process customer notifications. Messages can not only be transferred in written form and over the telephone, but also in electronic form, such as over the Internet.

Service Processing

Service orders represent all phases of the order processing starting with the planning of the activities, includes the execution, through to the completion of the service.

Deliveries of spare parts

A customer's spare parts requirements can be recorded using sales orders. The delivery is done on the basis of delivery notes.

Billing

Service products are assigned selling prices in the R/3 System. A billing request from the invoicing of the SD module can accept these prices automatically. A selection between different times and billing types can be made here. The order-related invoicing bills for costs resulting from internal activities, external activities and material resources are used for the billing. In contrast, for periodic invoicing, an invoicing plan is used for the individual service items or the complete service contract. The control over when and in which form an invoicing is to be made is determined using details for the start and end, periodicity (monthly, quarterly), duration and time of the invoicing.

Information system

The Service Information System, a component of the LIS, can be used to evaluate previous services. It provides a number of object-related and customer-related standard reports. Whereas the former covers the problem solution or execution of a service (service orders, service reports), the latter applies to information for sales and the shipping (customers, customer contracts).

4.5   The Customer Relationship Management-Initiative of SAP

With the Customer Relationship Management Initiative (CRM Initiative) SAP wants to expand a company's technological infrastructure to the customers. This places customer-related business processes in the focal point (Plattner 1999). The CRM Initiative groups new, existing and planned SAP functions into three new solutions:

- SAP Marketing
- SAP Sales
- SAP Service.

Both SAP Sales and SAP Service have their own mobile applications, which are called Mobile Sales and Mobile Service, respectively. The primary goal is to provide field service employees with mobile software solutions to simplify the support of the customers.

Whereas SAP Marketing is still being developed, SAP Sales and SAP Service are in the pilot phase. The following section introduces the three solutions and then discusses the architecture of the mobile software components.

4.5.1   SAP Marketing

SAP Marketing provides tools and functions for the planning, reporting and integration of marketing programs. The solution has the goal of providing segment analyses, database-supported market analyses and methods for market research. These analysis instruments can be used to identify target markets and to evaluate the effects of marketing programs. Users throughout the company can access both internal information and external sources, for example to market profiles from analysts.

4.5.2   SAP Sales

SAP Sales is a solution for sales support and makes use of some SD module components (refer to Section 4.3). The goal is to represent processes from the initial customer contact, through the order acquisition, and finishing with the delivery and service. Particular support is provided to field service employees with Mobile Sales, an SAP Sales component. The following section briefly describes the basic functionality:

- *Business Partner Management*: Field service employees and managers are provided with the capability to access central information about inquirers, customers and business partners.

- *Contact Persons*: This is an extendable reference for external, internal and private contacts. Field service employees can enter and maintain detailed information about contacts during their day-to-day work.

- *Activity Management and Calendar*: This component supports the sales during for the planning of appointments with customers and inquirers, visits by events and customers, telephone conversations and written actions. Information and statistics from the job reporting can be checked in regular intervals and made available throughout the company to other employees of the marketing team or other users. A calendar component provides a graphical overview of planned and outstanding activities arranged according to days, weeks or months as desired.

- Products and Services: This component allows access to all information needed to write a tender. In addition to the product and price information, for example, also graphs from the product or material master data.

- *Order Management*: This component makes available all data starting with the original customer contact through to the actual granting of the order, and provides a direct view of the sales channel. Field service employees can at any time view and obtain reports on the status of orders, product availability and various statistics.

- *My Infocenter (Marketing Encyclopedia)*: Advertising and sales documents, current news, competitor information or data from the Internet can be collected in a "marketing encyclopedia".

- *Price Determination*: As the name says, this component is used to calculate prices. For example, this enables sales employees to provide on the telephone immediate information about prices. The Sales Pricing Engine (SPE) uses the price and terms model from the R/3 System.

- *Product Configuration*: The Sales Configuration Engine (SCE) provides an interactive capability for a single-level and multi-level product configuration. This functionality is based on the variant configurator of the R/3 System, which is provided to the field service employee on his notebook. Because the SCE can work closely with the SPE, the corresponding price can be determined simultaneously for a product configuration.

- *Customer Agreements*: This component permits the access to the various types of customer-related agreements. These contain all binding agreements and business agreements signed by the customers.

4.5.3   SAP Service

SAP Service is used to coordinate all service functions such as customer support made on the telephone, spare parts delivery, service or invoicing. The Mobile Service mobile component is based on the SM module (refer to Section 4.4) and provides the following main functions:

- *Create and Dispatch Service Orders*: Service orders must first be created in the SM module before they are transferred to the notebooks of the service technicians. Several search and sort functions are provided in addition to all information that a service order contains, such as date and time, necessary equipment, customer information or on-site contact partners. In addition, all instructions, namely steps, that a technician needs to perform for the order can be requested.

- *Service Order Details*: The information offering that the service technician can display starts with customer master data with the associated contact partner, order history and the installed base, includes equipment master data with location and service history, through to contract data with information about existing service contracts, configurations, validity periods or supplementary terms.

- *Order Fulfillment*: After providing a service, a technician can enter the time for the journey and the repair. In addition, any problems that occur and the associated work can be logged and used for statistic reports, for example the frequency with which a specific problem occurs. Finally, work reports can be created and printed directly at the customer site.

4.5.4   Architecture and Connection of Mobile Components

Because the mobile applications for the SAP Sales and SAP Service solutions are fundamental components of the CRM initiative, this section briefly describes the architecture of these mobile components and the connection to the R/3 System.

The mobile applications are connected with the R/3 System using a Mobile Service Server (middleware). This server also serves as replicated database. The data are also distributed to local clients or to the central R/3 System or another system using keys, links and authorization concepts. Whereas the connection to the R/3 System permits a permanent data exchange, data matching with the clients is performed automatically in a defined cycle using a temporary online connection via telephone, radio or data network. To reduce the data transmission times, only those data fields changed since the last matching are transferred. Figure 4.1 shows the architecture.

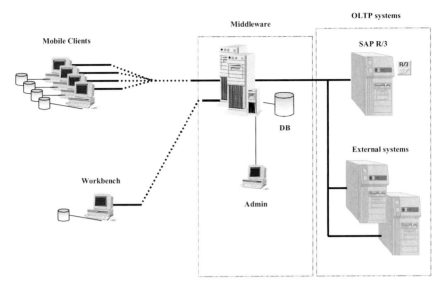

Figure 4.1: Architecture and connection of mobile components

In future, the functionality described in this section should be removed from the R/3 System and included in a self-contained CRM system. This also includes the "SAP Customer Interaction Center (CIC)" call center solution available with Version 4.5.

4.6 Electronic Data Interchange

4.6.1 Principles and Potential Use

EDI is understood to be the exchange of business documents between the computer systems from different companies. For example, an automobile manufacturer sends its orders electronically to its component suppliers or a telecommunications company uses EDI to send invoices to its customers.

To support the electronic exchange of documents, a range of industry-specific, company-specific and national standards have been developed. Examples of industry-specific standards are VDA (German Automobile Industry Association) in the automobile industry, SWIFT in the banking sector, SEDAS (standard rules for uniform data exchange systems) in the consumer goods industry, and DAKOSY in the transport sector. In contrast, national efforts produced the British TRADACOMS and the American Standard ANSI X12. The EDIFACT standard

(ISO 9735) developed by the United Nations, the European Union and the ISO standards organization exists as international standard since 1987. EDIFACT and ANSI X12 are currently the most popular standards, although their use varies greatly from country to country (Westarp, etc. 1999). A study made by the Institute for Information Systems at Frankfurt/Main University showed that 77 percent of the questioned German EDI users use EDIFACT, whereas in the USA just 32 percent use this standard compared with 64 percent who use the ANSI X12 standard.

The introduction of EDI initially results in costs that the US Chamber of Commerce estimates to average 50,000 US dollar. The main benefits are time and cost savings. The time savings result from the faster transfer of the data between the business partners and the automatic further processing in the operational software at the receiver side. In contrast to normal post, which can take several days, an EDI message takes often only a few hours or minutes to go from the sender to the recipient. The direct data transfer without media fragmentation accelerate the inter-organizational and inner-organizational processes. The wait times that result from the manual handling of a business process largely vanish. For example, there are no files in which orders must wait for their further processing, and also the manual acquisition of the incoming business documents becomes unnecessary. Forecasts estimate that approximate 70 to 95 percent of all data are created or entered more than once even though they are present in electronic form (Dearing 1990, page 5). EDI makes this duplicated input unnecessary. The time savings resulting from the automatic further processing of the EDI messages also affect the costs. The avoidance of the manual new inputs reduces the personnel costs. Other costs resulting from media fragmentation are also avoided. To these must be added the savings in postal charges for the traditional shipping of business documents and the costs for document management.

Despite these advantages, the use of EDI lies much below the forecasted expectations. This is principally because of the high implementation costs. Another important cost factor lies in the use of so-called Value Added Networks (VANs). These form the communications infrastructure for the exchange of EDI documents between business partners. Thus, it is not surprising that the use of EDI in small and mid-sized companies currently tends to be rather limited. For the future it is to be expected that EDI will be handled increasingly using the Internet. Section 4.3 discusses the possibilities and limits of these new forms of the EDI.

The application of EDI in the supply chain is manifold. For example, delivery schedules are sent using EDI, or customers receive their invoices electronically. As part of a case study, Chapter 6 discusses in detail the use of EDI in the supply chain.

4.6.2   EDI with SAP R/3: Possibilities and Limitations

SAP has developed so-called Intermediate Documents (IDOCs) to use EDI on the basis of the R/3 System. These are standardized descriptions of the structure of commercial documents, such as orders and invoices. This provides an interface to existing EDI standards. A conversion between the IDOCs and the EDI messages is needed to use EDI. To perform this translation, SAP-certified system providers offer EDI subsystems for the mapping, status tracking and the conversion of IDOCs into EDI standards, and in the reverse direction. Non-certified EDI subsystems can also be used.

As Figure 4.2 shows, each DOC consists of a control record, data records and status records.

DOC Record Types	
1. Control record	• DOC-ID • Sender ID • Receiver ID • DOC type and message • External structure
2. Data record	• DOC-ID • Sequence number • Segment • SATA
3. Status record	• DOC-ID • Status information

Figure 4.2: IDOC record types summary

The IDOC structures currently support messages that are comparable with the message types for EDIFACT or ANSI X12. The IDOC structure designations in Figures 4.3-4.6 make it obvious that SAP made particular use of the EDIFACT specifications.

Figures 4.3 and 4.4 show the EDI messages for the SD module supported by SAP R/3. A differentiation is made between incoming and outgoing messages.

Designation	From Release	IDOC Message	EDIFACT
Invoice	2.1	INVOIC	INVOIC
Quotation	2.1	QUOTES	QUOTES
Order Response	2.1	ORDERSP	ORDRSP
Delivery Notification	2.2	DESADV	DESADV

Figure 4.3: Outgoing messages from the SD module

Designation	From Release	IDOC Message	EDIFACT
Delivery Forecast	3.0	DELINS	DELFOR
Request for Quote	2.1	REQOTE	REQOTE
Order	2.1	ORDERS	ORDERS
JIT Delivery Schedule	3.0	DELINS	DELJIT
Invoice	2.1	INVOIC	INVOIC

Figure 4.4: Incoming messages into the SD module

The MM module supports as outgoing messages exactly the same message types as the SD module in the arrival. Similarly, because the incoming messages for the module MM are identical with the outgoing messages from the SD module, we can omit their representation. Figures 4.5 and 4.6 show the message types that the FI module supports for business finance.

Designation	From Release	IDOC Message	EDIFACT
Remittance Advice	3.0	REMADV	REMADV

Figure 4.5: Outgoing messages from the FI module

Designation	From Release	IDOC Message	EDIFACT
Invoice	2.1	INVOIC	INVOIC
Remittance Advice	3.0	REMADV	REMADV

Figure 4.6: Incoming messages into the FI module

Chapter 5   Internet Basic Technologies with SAP Systems

We have already mentioned at various places in this book the importance of the Internet for forming inter-organizational processes and the creation of new organization and cooperation forms. This chapter investigates the fundamental capabilities that the SAP R/3 System provides to support business processes over the Internet. These can be either business-to-business processes or business-to-consumer processes.

SAP provides three basic alternatives that make it possible to build and handle inter-organizational processes using the Internet (Pérez, etc., 1999):

1. The Internet Transaction Server (ITS) used to access the R/3 functionality. The ITS forms the interface between a Web-based application and the R/3 application server.
2. The Intelligent Terminal concept permits the direct communication to external applications using the SAP Graphical User Interface (SAPGUI).
3. Remote Function Call (RFC) can be used to access functions and data of the R/3 System.

Because the Business Application Programming Interfaces (BAPIs) are of special importance, the following section discusses them in more detail. Section 5.2 describes the use of the R/3 Internet Application Components and discusses their capabilities and limits. Section 5.3 then investigates the potentials and limits of the use of EDI over the Internet.

5.1   Principles of the SAP Business Framework: Business Components, Business Objects and BAPIs

The SAP Business Framework consists primarily of the following components (Pérez etc. 1999; Appelrath/Ritter 2000):

- Business Components
- Business Objects
- BAPIs.

Business Components are a defined encapsulated function repertoire that provide predefined interfaces. Examples for business components are Human Resource, Available to Promise, and Product Data Management. The mentioned examples already show that it is possible to differentiate between different types of business components, such as those which operate between application areas and Internet-relevant components (also refer to Section 5.2).

Business components can include several business objects. The goal of these business objects is to provide a standardized description of real-world objects, such as customers, invoices and orders. Business objects are described using static slots and dynamic methods. Those methods that are assigned to just one single business object are so-called Business Application Programming Interfaces (BAPIs). Figure 5.1 illustrates the business object "Customer Order".

BAPIs are realized as RFC-capable function modules. This provides an interesting opportunity to support Internet access to the R/3 System. For example, they can be used to create sales orders, or it is possible to read the product offering from the R/3 System, for example to generate a product catalog. The vision is that the developer only needs to know the interface definitions and do not need to know any implementation details.

Figure 5.1: BAPIs for the sales order business object

We consider the idea of the BAPIs to be basically suitable to make available a standard software solution using standard interfaces. However, because middleware components must be used, our tests have also shown that the external access to SAP systems using BAPIs and RFCs can be time-consuming.

Currently more than 1,000 different BAPIs exist, while the number increases with each version of the R/3 system. The BAPIs and business objects represent the business interface for external applications, on a logical level. In contrast, at the technical level, the integration can be made, for example, using HTTP, Java, COM/DCOM or CORBA.

5.2 SAP R/3 Internet Application Components

This section initially describes the Internet Application Components concept and shows selected applications. In particular, we discuss the FAG Kugelfischer application example in more detail. Finally, an assessment of the Internet Application Components is presented at the end of the section.

5.2.1   The Internet Application Components Concept

SAP, together with customers and external and internal experts, has developed so-called R/3 Internet applications to support Internet based business processes. These applications are called Internet Application Components (IACs). The Internet application components are software solutions to support business processes over the Internet.

The integration of the R/3 System with the WWW is performed using the SAP Internet Transaction Server (ITS), which administers all Internet-specific tasks of the R/3-Internet applications (Appelrath/Ritter 2000).

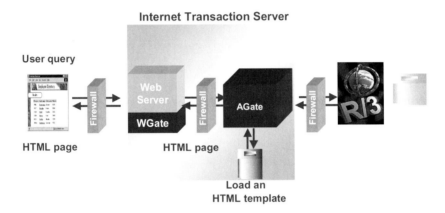

Figure 5.2: The ITS architecture

A major task of the ITS is the matching of the communication protocols and data formats between the WWW and the R/3 System. The ITS also performs a number of administrative tasks, such as Session Management and User Administration (Pérez, etc., 1999).

From a technical viewpoint, the ITS, as Figure 5.2 shows, consists of two components that can run on different computers: the Application Gateway (AGate) and the Web Gateway (WGate). The two gateways communicate using the TCP/IP protocol and can generally be used on different platforms (Appelrath/Ritter 2000).

5.2.2   Applications on the Basis of the SAP R/3 Internet Application Components

Version 4.5 of the R/3 System provides a total of 85 different application components (Internet Application Components).

The following section shows selected examples for the use of IACs. In the selected examples, we orient, in particular, on the capabilities for the support of logistics processes and the service management. The following Internet applications will be discussed:

- Online Store
- Sales Order Status
- Available to Promise
- Customer Service.

Online store application

The "*SAP Online Store*" application can generally be used both in the business-to-business and in the business-to-consumer area. This here is a classic electronic-shopping solution with integration into the R/3 System.

Whereas the order acceptance takes place in a WWW browser, the acquisition and the resulting order processing runs automatically in the R/3 System. Customers can choose products or parts, and then book them in the shopping basket of their procurement system (e.g. SAP Business to Business Procurement).

In the "*SAP Online Store*" application, the customer can order either products to be invoiced or pay immediately with a credit card. The complete payment process is based on SET (refer to the "The Use of Information and Communications Technology in the Processing Phase" subsection in Section 2.2.3).

Sales order status application

The "*Sales Order Status*" application enables customers to use their customer number and the corresponding password to view the processing status of their orders independent of time and location. Thereby, it is irrelevant whether one or more orders are present. After making the selection, an appropriate request to the SD module determines the processing status in the R/3 System before the data are forwarded to the inquirer.

Logistics service providers, in particular, provide their customers since several years with the possibility to use the Internet to track the transport of the deliveries. In addition, the computer manufacturer Dell, for example, also lets its customers use the Internet to request the current location of the ordered computer.

Available to Promise application

The *"Available to Promise"* (ATP) component is another interesting application that is generally suitable for both the business-to-consumer and the business-to-business areas. It makes it possible to quickly provide availability information.

The application permits, for example, customers to request information in real-time from the provider's R/3 System. After making the query, the application returns the binding details to the customers whether a specific product in the required quantity is available at the desired delivery date. The deliverable quantity is designated as the ATP quantity. If all the specified restrictions can be satisfied, the desired delivery date is confirmed.

When there is a positive query result, it is possible to enter the corresponding order directly into the R/3 System. In this case, the "Sales Order Entry" Internet Application Component is used.

Applications in the customer service area

Table 5.1 shows the various scenarios to support the customer service also contained in the IACs.

Scenario	Description
Create quality reports	Issue a certificate vouching for the quality of products
Input of quality messages	Entry of complains about defective of products
Input of service reports	Report a damage situation
Measured value and counter value acquisition	Enter measured values and counter levels, for example relating to leased plants
Query of the consignment inventory	Check the inventory of the supplier for the customer

Table 5.1: Scenarios to support the customer service

5.2.3   Electronic Commerce with SAP R/3: The FAG Kugelfischer Sample Application

This section uses the example of FAG Kugelfischer to show the use of IACs in a practical application. We also discuss the business assessment.

Short company profile

FAG Kugelfischer is the fourth largest supplier of anti-friction bearings worldwide with annual sales of approximately 3.2 billion DM. The company is organized into divisions and divided into company divisions:

- **Automobile Technology**: The Automobile Technology company division develops, manufactures and distributes worldwide anti-friction bearings in large-scale production, in particular for vehicles, motors and drive trains for motor vehicles.

- **OEM and Distribution Sector**: The OEM and Distribution Sector company division supplies worldwide standardized anti-friction bearings and customer bearings to customers from the focused industries of mechanical engineering, railed vehicles, steel industry, and the distribution sector.

- **Precision Bearings**: The production program of the Precision Bearings company division includes anti-friction bearings for the aircraft and space industry, and high-precision bearings and spindle units, mainly for the tool-making industry.

- **Sewing and Conveying Technology**: This company division comprises of the sewing and conveying technology business areas. The Sewing Technology division is concerned with the development, production and distribution of advanced special sewing machines, automated sewing machines and systems for the clothing, upholstery and automobile supplier industries. The conveying technology concentrates on the development, production and the marketing of conveying systems for light-weight goods and sorting plant for the distribution and the storage in the clothing and washing sector.

- **Components**: The production program of the Components company division covers forged parts, hardened inner and outer rings, ball-bearings, rollers, and sheet-metal and hard plastic cages for anti-friction bearings.

Current situation

FAG Kugelfischer makes use of a close cooperation with its trading partners to provide an indirect marketing channel. These partners sell the products to the end-customers. The communication for product requests, creation of offers, order of goods, etc., is mainly done by mail, by fax or telephone. Internally, the orders are booked into the SAP R/3 System and then further processed. Approximately 100 wholesalers have online access to the company's SAP R/3 System. This is done using a leased line with a SAPGUI front-end that is installed at these distribution partners.

Because FAG Kugelfischer would have to bear the cost for the providing the SAPGUI front-ends, the online access to Kugelfischer's system offered to the wholesalers is not economical for linking the many small or mid-sized companies in East Europe, Africa, Asia or South America. Furthermore, the smaller distribution partners would themselves have to pay high communications costs for the use of leased lines. In contrast, the costs for an Internet connection are estimated to be lower by a factor of 10. The distribution partners have the advantage that they can connect to a local provider and so can significantly reduce their communications costs.

Solution

FAG Kugelfischer can now use the Internet to provide another opportunity by which retailers can access the SAP R/3 System. The solution uses the ITS presented in Section 5.2.1. The following Internet scenarios were realized at FAG Kugelfischer using the new IT infrastructure:

- **Product catalog with sales order entry**: The retailer can view the product palette of 28,000 products in the Internet, and can order online the selected goods. The retailer has the option here of ordering at the specific date or immediately.

- **Order status**: The retailer can use the Internet to request the order status.

- **Report generation**: The retailer has the possibility to generate a report on the status of his order.

- **Availability check**: The retailer can check the availability of the goods before making an order.

- **Terms and conditions**: An inquiry for the specific terms and conditions for the product and price provides the handler with information on the current agreements with FAG Kugelfischer.

The significantly reduced costs in the mentioned scenarios mean the even small and mid-sized companies can have access to the R/3 System.

Project plan and costs

The project was organized into three main phases, namely the preparatory, realization and implementation phases. The solution concept was developed between October 1997 and January 1998, and realized in parallel on a test system (refer to Figure 5.3). The installation of the required hardware was completed in April 1998. Six months were required for the actual realization of the solution, namely the customization of the Internet scenarios (IACs).

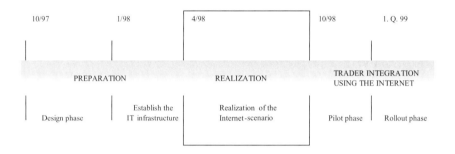

Figure 5.3: Project phases

Since October 1998, the first distribution partners could use the Internet to access the FAG Kugelfischer R/3 System. The solution with three distributors was successfully tested in an initial pilot phase. This meant that the actual rollout could start at the beginning of 1999. The rollout should be completed by the end of 2000 with the inclusion of 1500 distributors worldwide, which in turn have more than 10,000 branches.

The project cost for the solution implementation totaled DM 450,000. Of this, DM 100,000 was spent for hardware and DM 350,000 for consulting services.

Current results and benefits potentials

In accordance with FAG Kugelfischer's specifications, the use of the electronic commerce solution realized the following process improvements:

- The time-consuming information exchange between the distribution partners and the order control at FAG Kugelfischer to order goods no longer exists. The distributor can obtain all required information, such as prices, terms and conditions, and availability of the goods, from the Internet and use the same means to directly initiate an order. The previous written communication is no longer required.

- The average duration of the ordering process has reduced from approximately two days to less than 15 minutes.

- The implemented R/3 System solution supports or completely handles all process steps relevant for the ordering of goods.

Wholesalers that currently have access to the R/3 System using online access now have the option to use the Internet. This will avoid in future the costs for this online access that FAG Kugelfischer pays. There are onetime costs for FAG Kugelfischer of approximately DM 4,000 per wholesaler. They include the user license fees for the SAPGUI front-end, the costs for the hardware installed with the wholesaler, and the fees for the leased line and the R/3 Server. In contrast, the costs of an Internet connection is less than the online access by the factor ten. Furthermore, the previous training costs to learn to operate the R/3 user interface of approximately DM 2,000 per wholesaler are avoided. For the current number of 100 wholesalers, FAG Kugelfischer calculates a cost saving of approximately DM 550,000. Allowance must be made that this cost saving can naturally only be realized for the connection of a new business partner in comparison to the online solution using leased lines.

FAG Kugelfischer expects that the use of the solution will yield mid-term an annual sales increase of 25 percent in this market segment. There are two reasons here: there is improved customer involvement and the solution provides the chance to develop new business areas.

Another advantage for FAG Kugelfischer is the possibility to win distribution partners even in "emerging markets", which, because of the still unstable business structures, require only small volumes of goods. Because the solution greatly reduces the marketing costs for the company, even the order of small quantities is economical.

5.2.4   Internet Application Components – An Assessment

The presented applications illustrate that SAP is making an effort to provide support for inter-organizational business processes for important standard cases. The FAG Kugelfischer case study also shows the commercial potential that IACs provide. Many cases here are forms of the support that we no longer consider to be particularly innovative. For example, the online store described in Section 5.2.2 represents an electronic shopping solution that in the meantime has become commonplace in the WWW. Chapter 5 provides a more innovative form of electronic shopping solutions.

Thus, the advantage to the users of the IACs is not that they get a particularly innovative product, but they do not need to concern themselves with the interface to its internal processes, which, for example, are already defined in the R/3 System. This means that sales orders can be written directly into the R/3 System. Thus, the business logic specified for the R/3 implementation does not need to be redefined. For example, it is not necessary to reconsider the question which business partner, under which circumstances are to be offered which terms. Figure 5.4 uses the example of the Online Store to illustrate the integration between Internet applications and the R/3 System.

The FAG Kugelfischer application has shown how the use of Internet-based solutions can achieve significant commercial benefits in the form of cost and time savings.

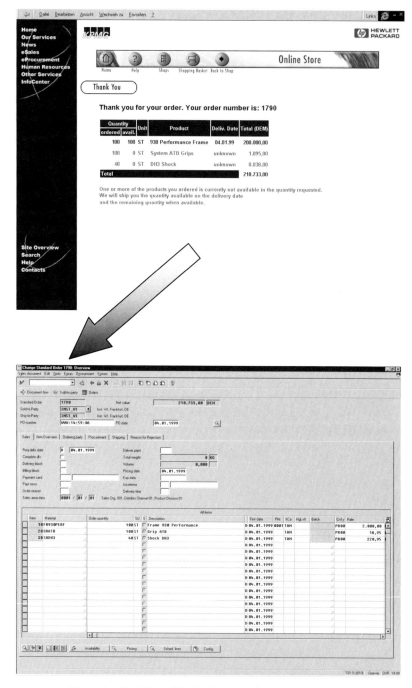

Figure 5.4: Booking of the IAC transaction in the SD module

5.3   EDI over the Internet

Section 4.6 has already discussed the importance of EDI and the capabilities and limits of EDI on the basis of the SAP R/3 System. This section now discusses the new opportunities of the use of EDI over the Internet.

In addition to the many advantages and "success stories" associated with traditional EDI projects, there is also a reverse side: currently the use of EDI is primarily restricted to large companies, and in a recent empirical survey, was mentioned as the most important reason for the implementation of EDI under pressure from large business partners (Westarp, etc., 1999). Forecasts show that only around five percent of the companies for which its use would be beneficial actually use EDI. In particular, small and mid-sized companies often mention the high implementation and operations costs also as reasons not to introduce EDI. Also, because many of the current solutions are platform-dependent, the use of EDI would cause additional investments for hardware and software.

The Internet with its wide availability, easy access and worldwide infrastructure could act as a catalyst for the diffusion of EDI, and provide a new impetus for EDI. The German EDI Company (DEDIG) differentiates here between WebEDI applications and Internet-EDI. WebEDI is understood to be the use of the WWW as basis for EDI applications. Existing solutions normally provide simple HTML input masks that permit a manual input of structured data. On the contrary, Internet-EDI designates the transport of EDI messages using Internet services, e.g. transmission using ftp (File Transfer Protocol) or smtp (Simple Mail Transfer Protocol).

Commercial advantages of WebEDI and Internet-EDI compared with traditional solutions result primarily through the reduction of the setup effort (implementation cost and manpower) and the operating costs. The use of open standards provides a platform independence, which can cause a significant reduction in the investment costs (Weitzel/Buxmann 2000). This gives rise to the hope that the entry barriers for small and mid-sized companies for the use of EDI networks can be significantly reduced. Consequently, the Internet satisfies almost all prerequisites for a replacement of traditional EDI solutions. Because the maximum transmission times for the use of the Internet cannot be guaranteed, there is a limitation only for extremely time-critical processes.

Because of the relatively open structure of the R/3 System, WebEDI and Internet-EDI solutions can be integrated into the R/3 System. For example, the transferred business documents can be written to the R/3 System using a BAPI.

SAP now provides an interface between the IDOC formats (refer to Section 4.6) and the Extensible Markup Language (XML). This can contribute to a more open form of EDI. In contrast to HTML, XML does not have a fixed format as a markup language, but rather is a meta-language that provides the rules to define an arbitrary number of concrete markup languages for a wide range of documents. This makes it possible to describe and structure data and documents so that they can be exchanged and further processed, in particular using the Internet, between a wide range of applications. Because the structure of a document defined in HTML cannot be determined, XML uses here the basic principle of separation of content, structure and layout (refer to Figure 5.5).

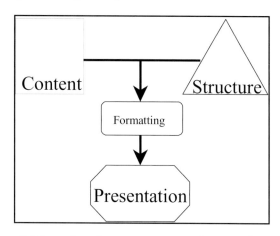

Figure 5.5 Separation of content, structure and presentation

A so-called Document Type Definition (DTD) that defines the syntax can be assigned to an XML document. This DTD describes the structure of a document type and defines which tags are mandatory and which are optional. The DTD can be used, for example, to check the validity of business documents.

In general, XML can be used to represent data structures of arbitrary complexity. For example, XML can also be used to represent and send EDIFACT message types. Consequently, XML appears to have the critical properties needed to realize EDI using the Internet.

```
<?XML VERSION="1.0" STANDALONE="yes"?>
<ORDERHEADER>
        <NAME>Smith</NAME>
        <FIRSTNAME>John</FIRSTNAME>
        <E-MAIL>Smith@anywhere.com</E-MAIL>
        <DATE>02.10.1998</DATE>
</ORDERHEADER>
<ORDERITEMS>
        <ITEM>
                <DESIGNATION>Harddisk</ DESIGNATION>
                <ARTICLENUMBER>123456</ARTICLENUMBER>
                <COUNT>5</COUNT>
        </ITEM>
        <ITEM>
                <DESIGNATION>Monitor</ DESIGNATION>
                <ARTICLENUMBER>9876</ARTICLENUMBER>
                <COUNT>1</COUNT>
        </ITEM>
</ORDERITEMS>
```

Figure 5.6: Sales order as XML document

Chapter 6 New SAP Systems to Support Supply Chain Management

The following sections describe new solutions to support supply chain management provided by SAP. These are

- the SAP Advanced Planner and Optimizer for the planning and optimization of logistic chains
- the SAP Logistics Execution System for the transport and warehouse optimization
- the SAP Business Information Warehouse that provides a comprehensive data warehouse solution
- the SAP Business-to-Business Procurement for handling procurement processes over the Internet.

The solutions are independent systems with their own R/3 independent release cycles.

6.1 SAP Advanced Planner and Optimizer (APO)

6.1.1 Overview

The Advanced Planner and Optimizer (APO) covers various logistics planning functions, which can be used both in conjunction with the R/3 System and also independent of it.

The APO supports the planning and optimization of a supply chain, starting with the ordering, through the production and assembly, and finishing with the transport. This is realized using several interacting modules that are supplied with data from OLTP or OLAP systems using interfaces. Figure 6.1 shows a summary of the APO components.

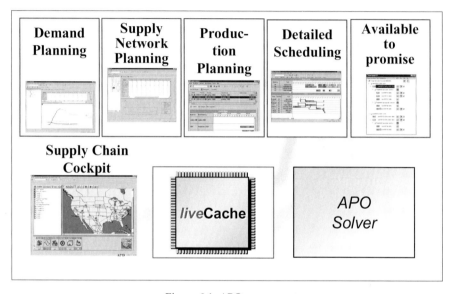

Figure 6.1: APO components

The APO components include

- the APO Solver that provides optimization algorithms and solution procedures for various logistics problems
- the Supply Chain Cockpit to display and monitor supply chains
- various planning modules
- a "liveCache technology" that constitutes a high-speed architecture for computing and data-intensive applications.

Thus, the APO consists of various components between which various interfaces exist. For example, the data storage in the "liveCache" serves as a basis for the use of planning and optimization algorithms. The following sections describe the APO components.

6.1.2   APO Solver

The APO Solver consists of a comprehensive library of optimization algorithms and methods for problems from the requirements, distributions and production planning.

For example, the APO for requirements planning contains simple methods from forecast modeling such as *exponential smoothing* and *regression analysis.*

To solve optimization problems in the area distribution and production planning *branch&bound procedures* (mixed integer linear programming), are used.

Moreover *Genetic Algorithms* (GA) are implemented in the APO. These imitate the biological evolution in order to find a best possible solution of an optimization problem.

The mathematical methods were developed to some extent on the basis of the ILOG Optimization Engines (http://www.ilog.com).

6.1.3   The Supply Chain Cockpit

The Supply Chain Cockpit (SCC) is a graphical user interface for the display, control and monitoring of a logistics network, and, in addition to the actual "cockpit", consists of a Supply Chain Engineer (SCE) and an Alert Monitor.

The user can navigate with mouse click through the system, and, when necessary, can fetch various information items from remote locations. For example, SCC can view and analyze warehouse inventory or individual sales orders. In addition to the use of APO applications and the information they contain, such as stocking and production schedules, the SCC also provides the capability of using the complete function repertoire of the Business Information Warehouse (refer to Section 6.3).

Figure 6.2 describes a geographical model using the SCC. The left-hand screen side displays all institutions defined within the model, for example production sites, distribution centers, suppliers and customers. It also provides information about materials and the transport between the individual locations. The right-hand screen side uses connection lines to illustrate the geographical positions of the locations and the transport connections between them.

The SCE, which creates application-specific representations in the SCC, supports the modeling and updating of network structures. This permits the graphical representation of networks of suppliers, production sites, facilities, distribution centers, transshipment locations and customers, namely the complete procurement and distribution channel. The SCE can also be used to define restrictions, e.g. specific goods can only be shipped with special transport equipment. For example, the planning modules of the system can use these conditions to automatically determine the transport route or the transport type. Restrictions modeled by key figures and attributes can be used in all APO subfunctions. Because the modeling of network structures using the SCE serves as basis for many planning activities, it is an elementary item in the use of the APO.

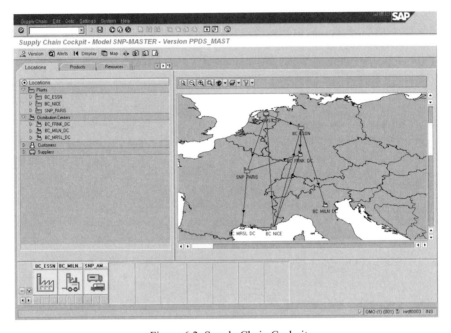

Figure 6.2: Supply Chain Cockpit

A further component of the SCC is an alarm monitor (*Alert Monitor*) that tracks various processes along a supply chain and automatically identifies a number of event-initiating problems and bottlenecks. A basic differentiation can be made in the APO between two forms of signaling: a status alert, which requires a user action, and a message alert, which only provides information without requiring user intervention. Figure 6.3 shows the use of the Alert Monitor to watch over the delivery dates.

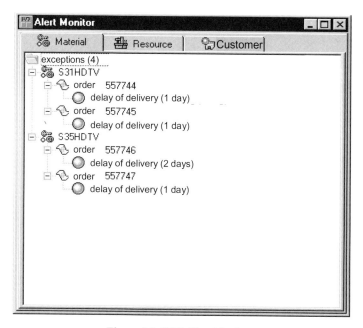

Figure 6.3: SCC Alert Monitor

6.1.4   APO Planning Module

The APO contains various modules to support the planning processes in the supply chain. These modules are for

- requirements planning
- logistics network planning (Supply Network Planning and Deployment)
- production planning and detailed planning
- the Available to Promise module.

The following section provides a short description of the functionalities of these modules.

Requirements planning

The APO contains several forecasting and planning tools for requirements planning and sales planning. APO can combine the use of these tools. These forecasting and planning methods can use data from various systems of different companies and from the Business Information Warehouse (refer to Section 6.3).

For example, the information originating from the operative systems for the daily work can be collected within the Business Information Warehouse and consolidated, and then passed to the planning. The operative data can be received from the SD, MM and PP modules of the R/3 System.

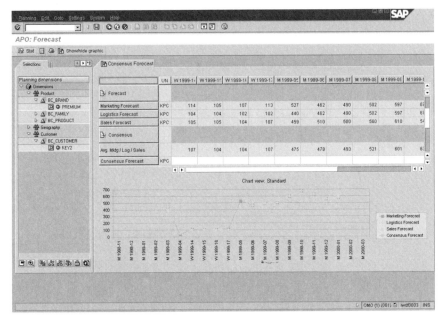

Figure 6.4: APO Demand/Sales planning

In addition to creating the common forecast, the module provides the opportunity to administer the life cycle of a product. For this purpose, information about the development of earlier, comparable products can be used. Promotional activities, for example short-term offers including advertising and their effects on the sales market, can be integrated in the planning similarly as for point-of-sales data.

Figure 6.4 uses a fictive example to show the requirement forecast in conjunction with prices, sales, advertising and the life cycle of a product.

Supply Network Planning and Deployment

This module can be used to plan inter-organizational interactions within a logistics network. The module uses the complete network relationships and their associated plans to provide plans for the areas of purchasing, inventory management and transport.

The *Supply Network Planning (SNP)* component acts between the sales planning and the capacity or production planning, and supports both areas. The user can define three different time horizons (short-term, mid-term and long-term). For example, a single "run" can determine all requirements deficiencies, requirements lead times and inventories. On the one hand, optimization methods such as the linear or mixed-integer programming, and on the other hand, SNP-heuristics, are used (Knolmayer, etc., 2000). This calculation results in a resupply plan, for example, for a selected material at a location of the complete model, using lot size profiles, receiving quotas, and the location distribution.

The *Deployment* component is an integral part of the logistics network planning and provides the capability to dynamically compensate and optimize a distribution network. The component determines which requirements can be fulfilled by the available reserves. If, for example, the available quantities cannot meet the requirements, the "fair-share strategy" is used to adapt the distribution schedule. This strategy is used to determine the current goods distribution using the available-to-deployment quantity, the outstanding sales orders, the safety stock, and the forecast. The goal is to

- distribute the inventory proportional to all distribution centers in accordance with the requirements
- increase the warehouse inventories in all distribution centers to approximately the same percentage of the target warehouse inventory.

If, however, there is an oversupply of a quantity or a lack of available warehouse space, push rules are used to compensate. If the planned quantities agree with the current plans, the SNP plan is confirmed. The combination of the SNP and Deployment components ensures an appropriate use of production, distribution and transport resources that caters for all restrictions within the supply chain.

After using the Deployment component, there is also the opportunity to use the Transport Load Builder, another component of the logistics network planning, to specify the loading of the transport while taking account of transport capacities and times. The loading is oriented on suggestions from the Deployment and groups the available products.

Production Planning and Detailed scheduling

The modules for the production and detailed planning contain a number of tools used to create production schedules while making allowance for any capacity restrictions. The specification of the production dates for orders and activities at the short-term planning level while taking account of general conditions, such as product and capacity availability, is called detailed planning. This component provides various functions to support the planning and decision-making process, and can be used to create an operative plan. The component covers

- a configurable planning board
- the automatic optimization of the planning with regard to various criteria such as minimum lead times, delays and set-up times
- simulations
- an automatic evaluation of the planning quality
- the display of planning conflicts and possible solutions.

The functionalities can be used for both individual locations and between plants.

The assignment of the production resources and the sequence planning is made either by a person or automatically using algorithms from the APO Solver (refer to Section 6.1.2). An order scheduling algorithm generates a plan taking account of available capacities and material. Starting with the due date, a backwards scheduling uses standard activity times to determine the start times for an activity. All parts of the bills-of-materials can be listed for the individual requirements. When material is available, it is assigned to the associated order as shown in Figure 6.5. The Alert Monitor indicates material that can be neither procured nor produced, and exceeded deadlines and capacity problems (refer to Figure 6.5).

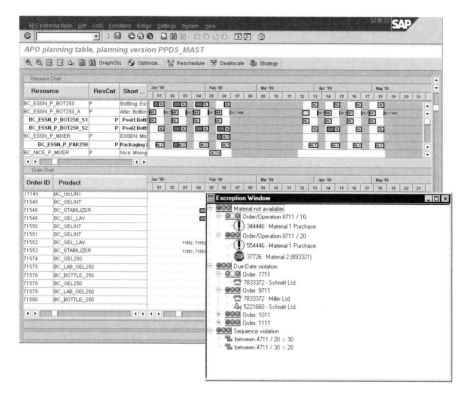

Figure 6.5: Production schedule with order network (including resources)

Example of an integration with the R/3 System:

An order request enters the SD module of a R/3 System from the Internet or EDI. The SD module initiates an ATP request for the products to the APO. The APO now determines that the products of the order must first be manufactured and forwards the order to the production and detailed planning. The available material and the available capacities are used to calculate an ATP date. This date is returned to the SD module. Because manufacturing bottlenecks automatically affect the delivery dates for sales orders, there is a close relationship between the sales and production orders.

Available to Promise

Available to Promise (ATP) is a functionality for multi-level checks of the resource and product availability, which permits an integrated, up-to-date availability test along a supply chain, in particular, for companies with a high throughput in sales and manufacturing. The ATP can be used to determine a delivery date or the effects of a desired date while making use of company internal and external areas and at the same time taking account of the costs, such as additional transport costs.

In addition to a pure availability test, the following capabilities are offered for alternative planning for bottlenecks:

- ***Product substitution***: If a product is not available, rule-based criteria can be used to determine an alternative product, which is either just as good or better.

- ***Selection of alternative locations:*** Should products not be available, it is possible to acquire the product from another location.

- ***Production***: When products are not available in sufficient quantities, this information can be used for the future production planning.

A rule-based operations planning can now be defined for these alternative planning methods. This determines, for example, in which cases and in which sequence the alternative methods are to be used.

The ATP also provides an explanatory and simulation component. It provides the capability to determine the effects of a new requirement for the product availability and interacts directly with the user.

The integration of all planning data within the liveCache (refer to the next subsection) makes the ATP information available in real-time and, through the connection to the R/3 System, results in an immediate update.

The SAP liveCache

The "liveCache technology" designates a high-speed infrastructure for complex, data-intensive applications within the SAP Business Framework. This was extended so that the execution and the associated data are more closely linked. All necessary information is loaded into the APO Server main memory and is processed there. The aim is to permit simulations, planning and optimization in real-time. The liveCache can be used by all APO components.

The liveCache technology uses the multi-processor and multi-computer configuration that permits a parallel and distributed processing. Currently, a maximum of three gigabytes memory can be addressed, although SAP is currently working on an extension to this restriction.

6.2   SAP Logistics Execution System (LES)

6.2.1   Overview

SAP Logistics Execution System (LES) is a system that supports a wide range of processes in the areas of distribution, transport and warehouse management. This is not a complete new system, but rather an extension and restructuring of existing R/3 functionalities, the basis of which lies in the *Warehouse Management System*. In addition, the LES is added with the name Logistics Execution (LE) at Release 4.5A to the component hierarchy of the R/3 system and so placed at the same level as SD and MM. LES, in particular, pursues the goal of better catering for the interactions between the areas of warehouse and transport management. The LES consists of two linked applications, the warehouse management system and the transport handling system. Because the Goodyear case study (refer to Chapter 8) provides a detailed description, the following section presents these two applications in limited detail.

6.2.2   SAP LES Warehouse Management System

The LES Warehouse Management System (WMS) extends the functionality of the Supply Chain Management to provide improved support for warehouse processes. These include the checking of the goods arrival, the specific individual material stock placement, the coordination of warehouse movements talking account of any special requirements, and the division or grouping of orders as part of the shipping handling.

WMS can be used to centrally administer different, even geographically separated, warehouse and production sites. It does not matter whether these are plant, central, regional, distribution or consignment warehouses. Both simple and complex warehouse structures can be represented. The assessment of the inventories in the warehouses is supported by common inventory methods such as fixed date inventory, sampling inventory, cycle counting and zero stock check. LES also contains several standardized queries to analyze the material flow. These include ABC analyses (refer to Section 2.1.3) of warehouse movements and summaries of all deliveries in a specific region. Meaningful reports, evaluation tools and ad-hoc queries can, for example, provide a warehouse manager with an early summary of the expected work load and so simplify the planning of the necessary resources. Because such queries require a very large amount of time and so cause performance losses, the liveCache technology should be used in future.

The facilities to support specific packing, marking and commissioning services are also new. In particular, LES permits a so-called wave-based, multi-level commissioning. Namely, the items or articles for different sales orders are arranged as commissioning waves that determine the internal processing in a warehouse. The picking is no longer made order-oriented, but depending on the overall volume for all orders. The internal activities then can be executed in that sequence that minimizes time and costs.

In addition to the goods input processes, LES also provides various functions to support goods issue processes. The shipping handling can be used to group various sales orders, subdivide large orders, and single orders or bulk orders optimized for delivery depending on the availability. This division can be performed automatically by the system on the basis of predefined rules or manually. All required shipping papers such as delivery notes, bills of lading or invoices can be printed to transmitted in the form of a shipping notification to the appropriate recipient.

As in the APO, an "alarm monitor" (*Warehouse Alert Monitor*) can also be used to supervise the process details. If processes do not finish on schedule or unexpected events occur, an alarm message is forwarded directly to the appropriate managers.

The LES is able to integrate modern, non-SAP technologies and to use these in the logistics handling. It provides a number of certified interfaces that permit these interactions, for example to technical devices such as "Automatic Storage and Retrieval Systems (ASRS)", Pick-to-Light systems and complex warehouse control computers. Of particular advantage for the support of large enterprises is the capability to link radio terminals using wireless data transmission. MOB from WITRON Corporation (also refer to Chapter 8) is an example of such a system.

6.2.3   SAP LES Transportation Management System

The *Transportation Management System* (TMS), the second component of SAP LES, can be used to plan incoming and outgoing transport, monitor their execution, and calculate the transport costs. The TMS does not provide any route optimization but just a simple assignment functionality.

However, TMS provides standardized interfaces for route optimization to a number of certified SAP partners. Queries from customers concerning the shipment planning and delivery scheduling agreed with the WMS provide binding information about the arrival of the shipment.

The TMS also provides support to calculate the freight cost and transport invoicing, which are closely related to the areas of financial accounting and controlling. Different freight tariffs, supplements and discounts or multi-dimensional scales can be used for this purpose. The calculation can be based on the weight, volumes, distance, transport mode, freight class factors or individual agreements and can be either one-dimensional or multi-dimensional.

A tracking is also supported that shows the current status of transport processes and permits a short-term response to plan variances. For example, not only loading activities, but also processes external to the own company can be continually analyzed. In future, the LES should provide support for so-called Global Positioning Systems (GPS) and Car Location Message Systems (CLMS). This connection makes it possible to continually track the location of transport means or deliveries (shipments), and permits monitoring between companies.

The TMS provides standard reports for the evaluation of data relating to the performed transport. There also exists an interface to the LIS of the R/3 System. This can be used to generate additional reports. For example, it is possible to determine how many deliveries a specific carrier performed in a certain period.

The integration of the management for warehouses and transport makes it possible to use the interaction between the two areas in the overall processing, the handling of which previously presented major difficulties.

6.3   SAP Business Information Warehouse (BW)

6.3.1   Overview

The SAP Business Information Warehouse (BW) is a comprehensive data warehouse solution that supports the representation and processing of large data volumes. The data can reside both within and external to the R/3 System. For example, company-external data can originate from suppliers, service suppliers or data providers. SAP has currently cooperation with the ACNielsen and Dun & Bradstreet data providers. The establishing of such a data basis forms the basis for standardized and customized evaluations and analyses. For example, the BW can be used to create so-called InfoCubes that permit a data analysis from various perspectives. Similarly, support is provided for the creation of reports, which can be used to make business decisions.

As part of supply chain management, the BW can use open interfaces, for example BAPIs, as common data basis for the participants in a supply chain. Thus, the BW can be used, for example, as basis for the application of the APO. Figure 6.6 shows the BW components.

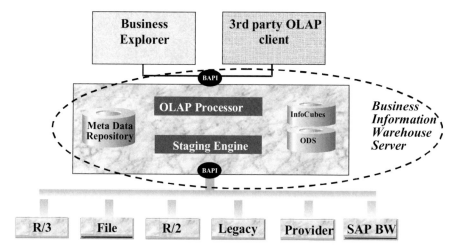

Figure 6.6: BW components

The following section introduces the techniques used to build the data basis before discussing the opportunities and limits of the support for the business decision-making process.

6.3.2   The Creation of a Data Basis for the Business Information Warehouse

Every data warehouse is based on a data basis that normally contains both current and historical information from all company divisions at various levels of aggregation. The Business Information Warehouse Server (BW Server) contains this data basis for the BW.

The BW Server provides extractors and transformation programs to create the data basis. The extractors are programs used to transfer internal and external company data and are specially designed to provide a high throughput of many parallel updates. They filter and consolidate the transaction and master data from the OLTP applications. The transformation programs represent the extracted data as data structures of the BW. The most important aspect here is the storage of the data with the aim of providing a flexible basis for analyses and reports. BW also contains so-called Staging BAPIs to load external data in cooperation with the certified partners ETI, Informatica, TSI and Prism Solutions.

To minimize the data volumes to be transferred to the BW during running operation, SAP provides two different procedures:

- *Timestamp*: Timestamps are assigned to the data records and those data records with timestamps more recent than the date of the last data transfer from the operative systems is selected and passed to the BW after making the transformation.
- *Change Log*: The data relevant for the BW is written during the processing of an OLTP transaction in their own log tables. BW imports these change logs at extraction-time.

All BW information is stored in a (meta-)repository. This is IT-relevant and business information, for example information about the base data model, the data origin, the performed transformation process, the available summary levels including the time-related processing, existing reports and analyses (refer here to the next section), and information in the form of a semantic description of all stored data. A range of catalogs contain various meta-information and classes:

- **InfoObject catalog:** The InfoObject catalog contains a description of all key figures and characteristics independent of their use.

- **InfoCube catalog:** This catalog describes the definitions of all InfoCubes with their key figures and characteristics.

- **Report catalog:** This catalog contains all report definitions and descriptions that can be displayed using a front-end. The catalog provides instructions to the OLAP processor which information is to be selected from an InfoCube.

- **InfoSource catalog:** This catalog stores all InfoSource definitions.

Such a repository often has the task to supply end-users with suitable tools to search for various data and information. The BW also provides the user with well-known instruments to obtain data from the system. These end-user front-ends are the Business Explorer used to display report data and also front-end tools from other manufacturers that communicate with the BW using the Microsoft standard interface OLE DB for OLAP. The Business Explorer Analyzer is embedded in Microsoft Excel.

6.3.3 Possibilities for Support of the Decision-making Process

A comprehensive data basis now provides various possibilities to support the decision-making process. All BW data is initially present in non-summarized form. The data from various source system are combined to a consistent format, which is designated as "InfoSource" (refer to Figure 6.7). The InfoCubes are then derived using summarization rules.

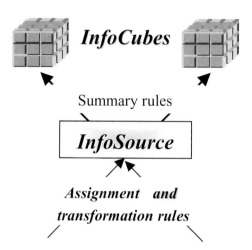

Figure 6.7: Derivation of InfoCubes

These InfoCubes consist of key figures and characteristics. The conceptional view for the continuously improved OLAP model covers business dimensions such as sales, costs, profits, quantities such as divisions and regions, and a time axis (refer to Figure 6.8) (Jarke, 2000).

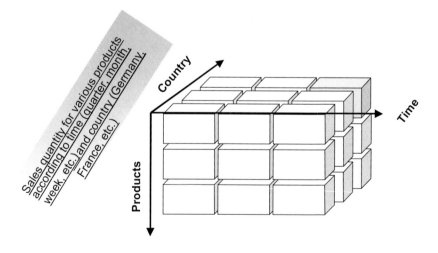

Figure 6.8: InfoCube

An InfoCube is formed from a set of relational database tables that are combined using a star scheme. The star scheme describes a center derived from a large fact table that is enclosed by several dimension tables (refer to Figure 6.9). The dimensions are joined using a foreign key relationship with one of the key fields of the facts table. The BW manages additional master data to the characteristics in the dimensions. The OLAP processor can be used to analyze the master data attributes, such as customer groups. A particular advantage is that the assignment of the characteristics attributes to this characteristic can be changed later without changing data or reorganizing the facts table. The facts table that stores key figures at the highest level of detail is used to connect independent dimensions with the key figures.

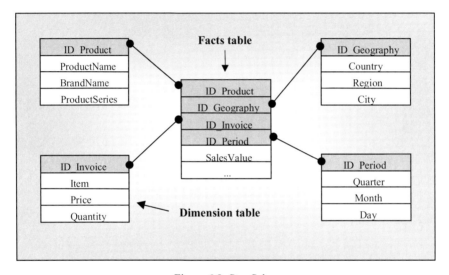

Figure 6.9: Star Scheme

BW also provides the opportunities to generate reports. BW stores the so-called workbooks on the BW Server for the company-wide access to the reports. The workbooks are managed in channels oriented on the user roles. The users can subscribe to individual channels and so receive requirement-oriented information. Figure 6.10 shows channels in a Business Explorer and the evaluation of a report in Microsoft Excel.

Every employee with access authorization can use the Business Explorer to request historical and current data at every desired level of detail. Because all reports can be stored in Excel, the data only need to be retrieved with the OLAP processor when an update appears to be necessary.

Figure 6.10: Request reports and information analysis

In contrast to most data warehouse solutions, the BW provides a business content (refer to Figure 6.11). This provides predefined models for analytical questions, appropriate extractors, display rules and requests for concrete reports that are based on the business processes of the R/3 System. These business models primarily contain decision objects, for example ordering, customer, supplier, material and plant, and key figures, which, for example, describe delivery reliability, capacity loading, lead time or order arrival for the logistics area. The decision objects and key figures are part of the R/3 business processes and can show, for example, country or industry-specific differences.

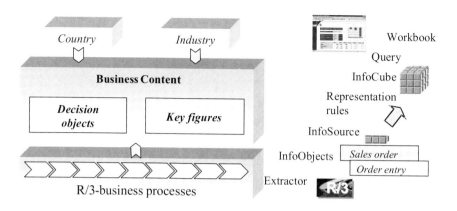

Figure 6.11: Formal model for the business content

The business content oriented on the information requirements of many customers, industries (for example automotive, service and transport) and countries offers the advantage that standard requests for various processes do not need to be programmed, because they are already supplied with the BW. Release 1.2A contains InfoSources for the logistics areas of sales, purchasing, inventory, production and project system, 18 InfoCubes, and approximately 30 queries.

As additional form of the decision-making support, the BW contains standardized key figures based on the recommendations of the Supply Chain Council (http://www.supply-chain.org).

6.4 SAP Business-to-Business Procurement

6.4.1 Overview

SAP Business-to-Business Procurement is a system used to support procurement processes that concentrate on C-materials and services. The handling of these processes from the requirements through to the payment is performed electronically using the Intranet or Internet infrastructure. Business-to-Business Procurement provides a Web-based user interface. The central initial screen designed for a wide-range of transactions permits employees from their workplace to purchase goods and services and check their price, availability and delivery time.

The direct connection to supplier catalogs is made from the viewpoint of the Business-to-Business Procurement solution using a so-called "SAP Open Catalog Interface". The application should support systems from Harbinger, Intershop, Requisite, and SAP (Online Store).

The Business-to-Business Procurement system is based on a three-stage client-server architecture. In contrast to the R/3 System, the access is not made using a SAPGUI but with a browser user interface that communicates with the Business-to-Business Procurement Server using the ITS. Release 1.0 B requires a connection to an R/3 application server in order to update or to create the associated subject matter in the MM and FI modules.

The SAP Business-to-Business Procurement 1.0 B solution consists of the following components:

- Internet Transaction Server 2.2

- Web Server

- Web-Gateway (for example MS Internet Information Server 3.0 and Netscape 3.0)

- Business-to-Business Procurement Application Server

- SAP Business Connector (BC) (from Release 2.0).

Business-to-Business Procurement also provides an SAP catalog that contains the Requisite catalog engine.

The support of SSL or SET takes account for the security aspects, such as authentication and encryption (refer to the subsection "The Use of Information and Communications Technology in the Processing Phase" in Section 2.2.3). So-called Procurement Cards can also be used to simplify the payment process.

In future, as part of the Business-to-Business Procurement, heterogeneous systems can be connected using XML. The basis here are BAPIs that encapsulate IDOCs and which the Business Connector translates into XML. This should make it easier in future to read and further process business content from different external systems.

6.4.2   The Procurement Process with SAP Business-to-Business Procurement

The starting point for the procurement process is a company's requirement for goods or services. SAP Business-to-Business Procurement can be used over a Web-interface to select various articles or services from a catalog system but also free entries, for example the material number. The solution supports various procedures of the catalog integration. These include the following capabilities:

- *Internal catalog connection:* Public company catalogs for purchasing are a widely-used form of the catalog connection. For this purpose, various catalogs are integrated in the internal systems of a company to support the employees' procurement activities. A bundled "SAP catalog" was integrated in the Business-to-Business Procurement solution to provide monitoring, security and performance capabilities for those companies that wish to use such a catalog in their intranet environment.

- *Integration of catalogs from different providers*: So-called "Multi-Supplier" catalogs provided by external brokers offer another capability. They support a wide range of different goods and services, and reduce the effort needed to make a free search.

- *External providers*: The Business-to-Business Procurement architecture can also be used to directly access external catalogs using the Internet.

These catalogs permit SAP Business-to-Business Procurement to make automatic product and price comparisons, and to check the availability. Currently, an availability check can be performed only when the products are offered over the SAP Online Store (refer to Section 5.2.2).

Because users can be assigned in activity groups, e.g. employee, manager, secretary, these are allowed to order various articles depending on their authorization without requiring the approval of a manager. When a requirement needs the approval of a manager, the requirement is forwarded to her or him using a workflow. If the requirement does not need any approval or when this already has been performed, a purchase order, a reservation, a requirement request, or a service entry sheet is created in the R/3 System. The application now completes and verifies all required supplier data. The master data for the employee who initiated the requirement coverage request, for example his cost center, are automatically added using the user masterdata, and permit queries at any time.

Figure 6.12 shows the system architecture needed for an interaction with the R/3 System. An employee's ordering activity performed using a Web-browser uses the

EC-application to reach an R/3 for further processing. Orders created in the MM module can be transferred using existing EDI-interfaces, e-mail, fax or the Web to the providers or suppliers. The figure contains the BC middleware first available with Version 2.0 that permits a conversion into XML. It permits future messages to be transferred to the vendor page.

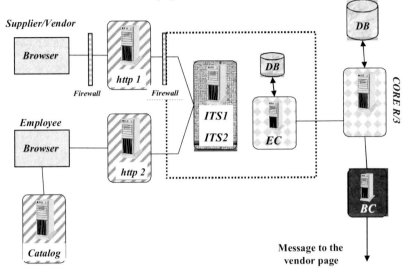

Figure 6.12: Interactions between systems and transaction partners

Given the appropriate authorization, invoice and performed service information for the vendor page can be directly entered into the Business-to-Business Procurement system using a Web-interface. This function replaces the traditional paper trail. The associated payment process is initiated after it has been checked and confirmed by the ordering party. When services are involved, the Web-interface can be used to add all necessary service information to the applicant's system by the vendor (in this case a so called service entry sheet is created).

Figure 6.13 uses an example to provide an overview of all order requests and material sales.

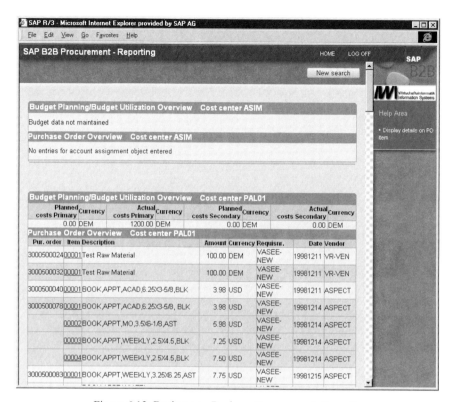

Figure 6.13: Business-to-Business Procurement: Reporting

To obtain more detailed analyses, an integration with the BW has been announced for the Business-to-Business Procurement solution Release 2.0 B (general availability, in the middle of 2000).

6.5 New SAP Systems: Status quo, Experience of the Use and Perspectives

The systems presented in this chapter are still relatively new and, to some extent, not yet established in practical use. Because this is a dynamic environment in which rapid and extensive changes occur frequently, we have created a moderated Web-page at http://www.wiwi.uni-frankfurt.de/sap that documents experience with the use of the systems described here. Readers can apply for a password that permit participation on the network.

The *Advanced Planner and Optimizer* has been supplied to a total of 15 pilot customers and has been available to general use since December 1998. In this first version, the APO supports only Oracle databases and the Microsoft Windows NT platform. Worldwide, currently approximately 130 installations (Release 1.1) have been delivered. Release 2.0 also accepts UNIX as platform. Our Goodyear case study describes first user experience and results of a feasibility study. With its comprehensive method base, the APO provides a good perspective for the optimization of processes along a supply chain. In particular, together with the BW it can be used for the cooperative decision support.

The *Logistics Execution System* was implemented under the name Logistics Execution as Release 4.5A in the component hierarchy of the R/3 System and thus supplied to all customer of the R/3 System from Release 4.5. The Goodyear (Chapter 8) case study shows the opportunities of the LES application for the management of transport and warehouse processes. The functionality has currently been used by more than 1,000 companies.

Since April 1999, Version 1.2B of the *Business Information Warehouse* is available. Compared with the first generally available Version 1.2A, this version provides several functional extensions, a number of performance improvements and an extended business content. The Schenker Logistics and Goodyear case studies show the opportunities and limits of the BW application. Approximately 600 companies currently use the BW.

Version 1.0B of the Business-to-Business Procurement generally available since April 1999 supports the 4.6.B, 4.5B, 4.0B, 3.1H and 3.1I versions of R/3. The experience gained worldwide and over various industries during the pilot phase of the alpha version (1.0A) with regard to performance, ergonomics, etc., has been included in Release 1.0B. The release of Version 2.0B is general available in the middle of 2000 and supports also non-R/3 systems. SAP has announced that this solution will contain a so-called "Lean MM". This removes the connection to the MM module currently required. This means that orders can be created directly from the Business-to-Business Procurement system. The consequence is that master data, such as suppliers and documents for orders, are no longer exclusively stored in the R/3 System (as in Business-to-Business Procurement 1.0) but are also locally stored in the Business-to-Business Procurement system. Using a middleware these master data can be replicated from the backend systems. Because we do not discuss the practical use of Business-to-Business Procurement in this book, we again refer to reader to the above mentioned Web-page for current information.

Chapter 7   Schenker Logistics Services Case Studies: Management of a Supply Chain for Automobile Manufacturing Overseas

Schenker Corporation is one of the leading international suppliers of integrated logistics and transportation services. The company located in Essen, Germany and part of Stinnes Corporation has approximately 27,000 employees at almost 1,000 locations. It is organized into the company divisions *Schenker-BTL,* the largest logistics and transport company in Europe (specializing in European land-based transport), *Schenker International* (worldwide transport solutions in air and sea freight), and *Schenker Logistics*, which is considered as part of the case studies presented in this chapter.

Schenker Logistics is concerned with an area that range from complex warehouse systems through to additional supplementary services. Schenker uses the Supply Chain Management to provide potential customers with various system services that consist of components that can be combined in a flexible manner (*base services*). A major focal point of these services is the development and provision of complex IT solutions. Schenker concentrates here on the development of systems based on the SAP R/3 standard software. Since mid-1998 Schenker is official development partner of SAP in area "*SAP Automotive*".

In the first case study, we investigate in Section 7.1 the *status quo* of the IT-usage to support a supply chain. A second case study analyzes the business opportunities of the Business Information Warehouse.

7.1   Management of a Supply Chain for Automobile Manufacturing Overseas – Status quo

7.1.1   Overview of the Case Study

Our case study examines the management of a supply chain for the production of an A Class model by Mercedes-Benz in Brazil. The factory in Juiz de Fora ensures the supply to the South American markets. Schenker Logistics undertakes a large part of the organization, structure and monitoring of the information and goods flows accompanying the supply chain that ranges from Europe to South America. Figure 7.1 provides a simplified representation of this process.

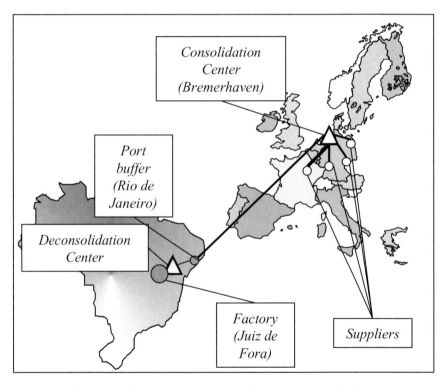

Figure 7.1: Geographical overview of the complete supply chain

Goods required for the production in Brazil form the initiating impulse for the supply chain. Suppliers initially deliver this material to a *Consolidation Center (CC)* in Bremerhaven (Germany). This CC is a concentration point for goods and information to which specific parts are supplied for the overseas transport to

Brazil. They are then packed, placed in containers and shipped to South America. In exceptional cases, aircrafts are used for the transport. When the parts arrive in Brazil, they are delivered to a so-called *Deconsolidation Center (DC)*. This can be considered as being the opposite pole to the CC in Bremerhaven. The DC has the function of a dissolution point in which the delivered containers are received and unpacked, and custom duties paid for materials and parts. From there, they are supplied just-in-time to the factory. The DC has the task to provide the consolidated material in the immediate vicinity of the "*point of use*" for the assembly or the production of the customer. It is immaterial here whether the assembly parts are supplied by sea, air or land.

Schenker Logistics operates the CC in Bremerhaven and the DC in Juiz de Fora. Both are transshipment centers for goods and information.

The following section describes the complete physical and information-logistical infrastructures, and the activities and processes within and between the participating companies in the supply chain. In particular, we concentrate on the functionality of the R/3 System used for intra-organizational and inter-organizational processing. The most important operative system activities and the required interfaces and data exchange formats for inter-organizational cooperation are discussed and explained. The description is oriented on Figure 7.2 and is subdivided into four different areas.

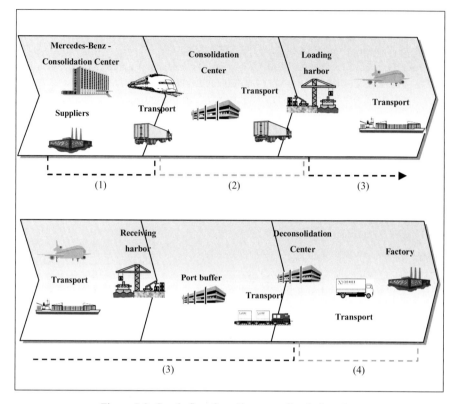

Figure 7.2: Goods flow from Europe to South America

Whereas the first considered area covers orders with the supplier through to the goods arrival in the CC, area two contains all necessary measures and services through to the loading of the parts for the overseas transport at the port of loading. Both areas represent the European part of the supply chain.

The physical loading in the harbor starts area three, and covers all required processes until the parts arrive at the DC in Brazil. Area four follows with the delivery to the factory for the A Class in Brazil. This also completes the supply chain. The last two subareas describe the South America view of the supply chain.

7.1.2   Management of the Processes to Supply the Consolidation Center in Bremerhaven

The starting point for all activities accompanying the considered supply chain is the Mercedes-Benz (MB) factory in Brazil, which passed its requirements for the production to a Mercedes-Benz Consolidation Center (MBCC) located in

Böblingen near Stuttgart (Germany). The first task of this MBCC is to receive, monitor and further process all the data required for the process along the supply chain. The information received from Brazil contains the delivery schedule needed for the production of the A Class and forms the basis for all following activities and processes. The requirements are based on the rough-cut planning that covers a period of six months. These requirements are later made more specific using a JiT schedule.

Delivery schedules are sent as EDI message in VDA.4905 format, converted with an MBCC interface, and processed using the SD module of the R/3 System. A sales order automatically creates an order request including delivery dates and delivery quantities. This forms the request for the purchasing department to provide the necessary materials, parts and services in the specified quantity and at a specific date.

In the further processing, purchase requisitions are automatically converted into orders and produce a scheduling agreement or delivery schedule for the suppliers. These delivery schedules form the basis for the data records transmitted to all suppliers. This data originated from Brazil is passed on in the same format (VDA.4905). As mentioned previously, this occurs six months prior to the actual production in Brazil and provides suppliers with the capability to match their production to the customer's demand.

The transferred data contains delivery date, article, quantity and, if necessary, other information needed by the suppliers. Because the MBCC can query the inventory levels of the Schenker CC in Bremerhaven, it is possible to match the demand with the available inventory levels. If, for example, there is large inventory for a specific product in Bremerhaven, correspondingly smaller subquantities can be requested from the suppliers.

The data transferred as scheduling agreement to the suppliers is also transmitted to the CC in Bremerhaven managed by Schenker. As the screenshot in Figure 7.3 shows, the MM module imports these data and uses them as basis for the planning and monitoring processes.

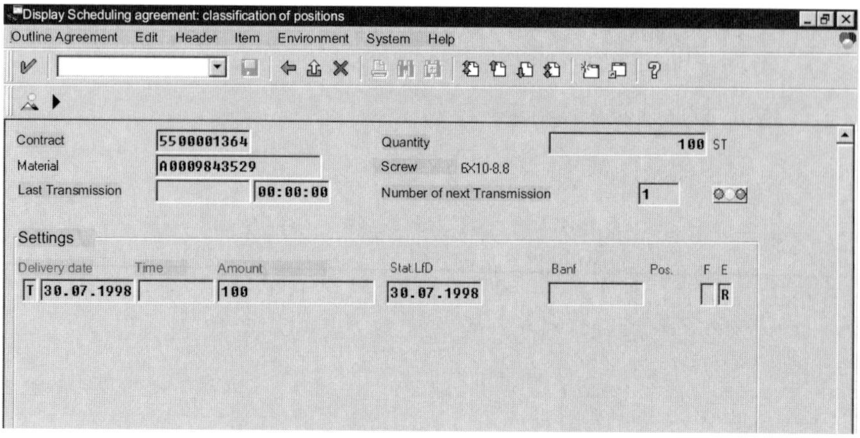

Figure 7.3: Transferred scheduling agreement in the MM module of the CC

As Figure 7.3 shows, the data from a scheduling agreement indicate that the arrival of one hundred screws of the material A0009843529 can be expected on 30.07.1998. Schenker can use this information from the scheduling agreement to check notified and future arriving goods for inconsistencies.

The data is generated by the R/3 System of the MBCC using modified IDOC messages (an extension of the DELFOR.01 Standard) and imported into the Schenker system using an appropriately modified IDOC interface.

The suppliers and regional service agents used by Mercedes-Benz also pass their data in the form of a delivery notification or shipping notification directly to the Schenker Logistics' R/3 System. This data is used as a basis for the delivery of parts by the associated regional service agents, which orient themselves on the previously transferred time windows and unloading points. Trucks are used to transport the various goods to the MBCC. In addition to the previously described tasks, the MBCC also acts as supplier. Thus, all components produced by Mercedes-Benz, such as materials from a factory located in Rastatt (Germany), are coordinated by the MBCC and delivered by rail to the CC (which has its own siding).

7.1.3   Management of the Processes from the Goods Receipt in the CC Bremerhaven Through to the Loading in the Shipping Harbor

Processes after goods receipt

Once goods arrive in the CC Bremerhaven there is an immediate comparison of the notified and delivered material, for which there are currently around 600 different parts. The goods receipt acquisition consists of a number of validation measures, which cover the control of the packing, the material and the provided data and documents. The CC can use this data and the previously received information any time to check for discrepancies and so localize problems. Figure 7.4 provides an overview of the checking measures.

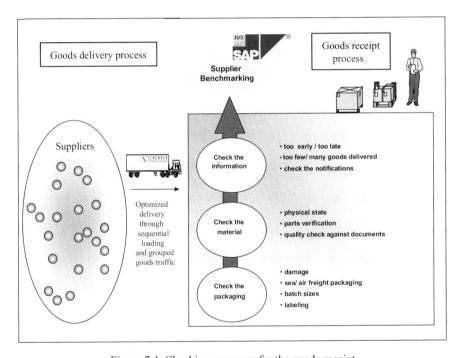

Figure 7.4: Checking measures for the goods receipt

This makes it possible, to perform an alternative planning in case of discrepancies. Thus, to avoid delays in the supply chain, long-delayed deliveries could be transported to Brazil by aircraft rather than by ship.

Schenker Logistics determines the service level (Benchmarking) of all suppliers (refer to Section 2.4). The analysis is oriented on the tasks fulfillment and is stored in the R/3 System. The standard functionality of the MM module is used here (refer to Section 4.2). This permits all suppliers to be assessed using standardized criteria. This evaluation helps the purchasing department of the MBCC and affects the supplier portfolio accordingly.

After the data and the physical goods arrivals have been checked, these are updated using the input mask shown in Figure 7.5.

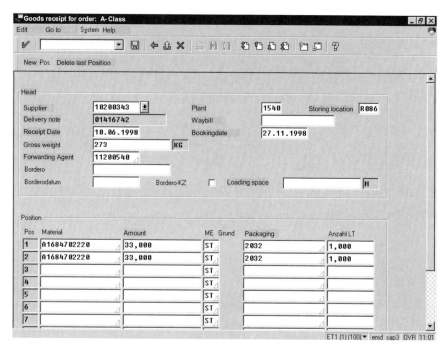

Figure 7.5: Goods receipt in the MM module

This is used to record information about suppliers, service agent, delivered materials and packing in the R/3 System. The updated data records are automatically transferred with a goods receipt message to the R/3 System of the MBCC where they are processed in the MM module. The transfer takes place hourly provided data records to be transferred have accrued. An IDOC format is used as basis for the data exchange. This format is based on SAP's planned standard format (DESADV.01). However, because this is an extension of the standard format, the data transmission is performed using an interface specially developed by Schenker. The actually delivered and canceled goods receipts are read using the goods receipt function of the R/3 System for MBCC. Depending on the situation, additional messages can be transferred about returns to suppliers, scrapping, stock quantities or inventory differences (*with reference to the party that caused the discrepancy*). The performed data transfer causes a redundancy in the two involved materials management systems for Schenker Logistics and Mercedes-Benz.

In addition to the goods receipt check, the goods receipt in Bremerhaven causes further processes to be initiated that depend on the loading equipment used, such as pallets and containers, and the packaging of the suppliers. All incoming parts initially represent a non-dispatchable inventory for the goods receipt. These parts must be subjected to various measures to change them into a dispatchable state. As Figure 7.6 shows, different procedures are possible depending on the type of the delivery:

1. **Cross Docking:** When the suppliers pack the parts in the form of end modules, no additional measures other than marking are necessary. The end modules can go without interruption into the dispatching inventory or to the loading.

2. **Reboxing:** Incoming parts are separated from their racks or crates by reboxing and are secured or stored on new racks or crates that are more suitable for handling.

3. **Repacking:** The parts shipped by the service agents are accepted and given special packaging before they can be forwarded. For example, this can be a crate with 5,000 items that is to be repacked into ten crates each with 500 items.

4. **Repacking & Reboxing:** This here is a combination of cases two and three. In this two-stage process, the incoming parts are separated, relabeled and commissioned differently, and packed into appropriate batch sizes before transition can be made to the dispatchable inventory.

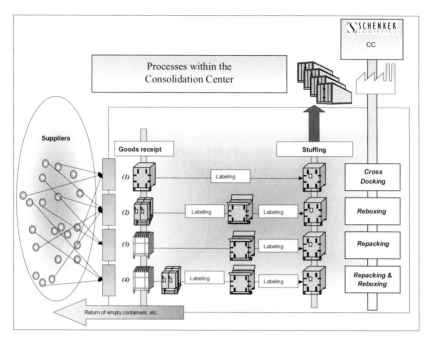

Figure 7.6: Internal processes of the CC in Bremerhaven

Because the R/3 System does not provide any functionality for the optimized stocking of loading equipment and containers (refer to the "Storage Space Planning" subsection in Section 2.1.3 for methods), the CC uses external software. This is a Fraunhofer Institute application the with the name "PUZZEL".

The CC uses the LIS to manage both full and empty loading equipment within the supply chain. Schenker Logistics produced the software solution needed here.

Goods issue processes

Because Schenker needs not only scheduling agreements but also the appropriate information on concrete delivery dates for the goods receipt, there is a further data transfer from the MBCC to the Schenker system in Bremerhaven. The data originating from the SD module for the MBCC is transferred in an extended DELFOR.01 format using a non-standardized IDOC interface into Schenker's R/3 System. The data is processed using the SD module and define from which date or from which cumulative delivered quantity a change status is to be sent to MB Brazil. The incoming delivery dates cause an automatic dispatch scheduling to be performed. The maintaining of the delivery dates takes highest priority here.

Figure 7.7 shows a scheduling agreement sent from the MBCC to the R/3 System of the CC. This agreement contains planning data that specify when and in which quantity a specific material must arrive in Brazil. Each scheduling agreement applies to a single material and can contain different delivery dates (these are 16.09.1998 and 03.11.1998 in Figure 7.7). Additional time restrictions, such as the delivery time to the factory, are not taken into consideration yet.

Once all consolidation activities within the CC have completed, the containers are transported to the loading harbor and loaded onto the planned ship. This is represented in the SD module with the booking of the goods issued for the associated delivery. The material removals are deducted from the stocks in the MM module and the status within of the R/3 System updated correspondingly.

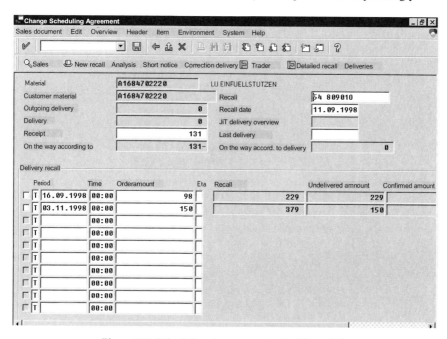

Figure 7.7: Scheduling Agreement in the SD module

The physical loading of the containers on the overseas transport medium (ship/aircraft) completes the European side of the business transaction and initiates another data transfer. The loading messages, which contain the information on the contents of the containers loaded in Bremerhaven, are forwarded as a so-called Advanced Shipping Notification (ASN) to the MBCC. It contains a message about the goods issue. In case relevant data has accrued, the ASN is send in a sequence of fifteen minutes. In analogy to the previously described modifications, an interface created by Schenker and based on the DESADV.01 standard data format is used here.

The information for the loaded containers is also forwarded as ASN to the DC in Brazil. In contrast to the previously described data transmission to the MBCC, no modification is needed. Because the IDOC data format corresponds to the standards (DESADV.01) for shipping notifications, the transmission is made using standard interfaces. Additional information, for example about damage or supplier details, is not required because it is not needed for the further processing by the DC.

The ASN also defines the transition of the processing to the responsibility of the DC in Brazil and so completes all transactions at the European side.

Support for finance processes

For the payment of the service agents, freight data accepted by the transport companies after being checked by Schenker is forwarded to Bremen using the dedicated file interface of the R/3 System. The internal Daimler-Benz format (Inhouse T1499A) used here is received and processed in Bremen with a corresponding interface. The data, which is transferred daily, is used for the invoice verification or to pay liabilities with the service agents and differs in their purpose from those already forwarded to the MBCC in Stuttgart.

7.1.4   Management of the Processes from Loading Through to the Goods Receipt at the Deconsolidation Center in Brazil

The ASN produced in Bremerhaven provides the DC with all information necessary relating to the transport medium and the containers it holds. It also contains the expected arrival date in the harbor, which is designated as "Estimated Time of Arrival" (ETA). Figure 7.8 illustrates this with a listing of all notified containers within the R/3 System of the DC. This shows that in addition to the containers arriving with the ship "Cap Polonio", two items are also transported as air cargo to Brazil.

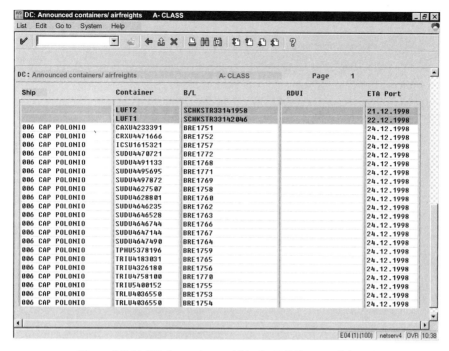

Figure 7.8: Notified containers within the R/3 System of the DC

This listing provides the DC, the European "counterpart" of the CC, with a control function over the Brazilian area of the supply chain in order to recognize any potential delays and to initiate the appropriate alternative measures.

After the overseas transport and the unloading of the containers in the destination harbor, these containers are transported without delay with trucks to the customs-free intermediate warehouse "*Port buffer*" within the harbor (Rio de Janeiro). This part of the supply chain prepares the containers for the further transport to the DC. The goods receipt in the Port buffer initiates the associated booking that is forwarded as goods receipt message to the DC system.

This message is transferred as a pure file transfer and so is imported into R/3 System of the DC using a file interface. Starting at this time, the supplied parts are maintained as inventory in the MM module. The transfer provides the DC with the information it requires about the arrival harbor, the ship used, the arrival date and the number of the incoming containers, and so provides the capability of monitoring the Brazilian part of the supply chain. The booking is performed using the previously received shipping notifications. This permits the control of goods receipts and tracking of any discrepancies.

Once the goods receipt booking in the port buffer and the approval for the transport without being subject to customs dues have been made, all containers are loaded onto trains and sent to the DC. This is located in the immediate vicinity of the factory and, as with the CC, also has its own siding. Compared with truck transport, railway transport has the advantage of more precise arrival times and lower costs.

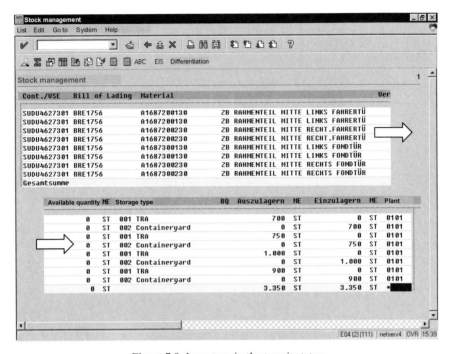

Figure 7.9: Inventory in the transit status

In case of time-critical situations the transport is performed with trucks. This achieves a time advantage of approximately sixty percent. Once the containers have been loaded, a new file transfer is made to the Port buffer (goods issue message). This is performed in a similar manner as the previously described transfer of the goods receipt. The corresponding files contain information such as the train number and result in an update of the status in the R/3 System of the DC. This goods issue message causes the containers maintained in the port buffer warehouse type to change to a transit status and to remain there until they arrive in the dry port or container yard. Figure 7.9 illustrates this activity in the R/3 System of the DC. Because no access to the containers can be made during the transport, the available quantity within of the system is reset to zero until the goods arrive in the container yard. This container warehouse is part of the DC (refer to Figure 7.10) and is located in the immediate vicinity of the actual warehouse in which the containers are emptied.

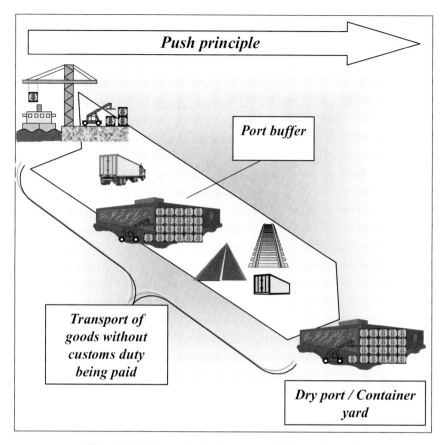

Figure 7.10: The push principle as part of the supply chain

Figure 7.10 illustrates the previously described route of the containers from the acceptance at the harbor via the intermediate warehouse through to the container warehouse. All containers coming from Bremerhaven are recorded and "pressed through" to the container warehouse of the DC using the push principle. Up to the time of the goods arrival in the container warehouse, the containers and the contained parts are not subject to customs duties. To save costs, customs duties are paid on the goods immediately before they are used. The system status is again updated after the goods receipt in the container warehouse. Namely, the containers are removed from the transit status, which describes an interim status of the containers between den two warehouse types port buffer and container yard, and entered in the DC container warehouse.

7.1.5   Management of the Processes from the Transition of the Containers in the Warehouse Through the Delivery at the Factory

The MB Brazil factory now passes additional data to Schenker that fix the rough-cut planning previously forwarded to the CC and specify the parts that are required within the next 11 days at the factory. Thus, containers are no longer handled but rather a resolved requirement for parts based on the production figures for vehicles.

The detailed requirements are transferred to the DC using a typical standard data format for Brazil. The data format used here is an EDI standard that is closely related to the previously discussed VDA format. An external EDI converter transforms the data into the R/3-IDOC format for shipping notifications (DELFOR.01). On arrival of the Scheduling Agreement a MRP is started to check if the requirements of the Mercedes-Benz factory can be fulfilled using available stocks.

All displayed warehouse types are managed centrally within a system. The MRP system checks the defined stocks and, should there be insufficient inventory, creates receipt elements that have the form of an purchase requisition in the area of the materials management system. Thus, to assure on-time delivery of Mercedes-Benz, material is required from the container warehouse of the DC or from earlier warehouses in the supply chain. Note here when a container is opened, customs duties must be paid for the complete contents or for a complete container delivery. Because the container warehouse is stocked with many containers and various materials in different quantities, the performed scheduling agreement updates may mean that it is desirable to extract a different container than that originally planned for a specific period. The availability check is made through all warehouse levels at the Brazilian side that are assigned in the R/3 System with different "location numbers" (refer to Figure 7.11) for each warehouse type. If the required parts are not present in the container warehouses, a test is made whether they are available in the transit status or in the intermediate warehouse at the harbor. This additional service makes it possible to delay paying customs duties until an optimum time. Because the customs duties for the parts can be associated with some avoidable costs in the form of an unnecessary tying up of capital, the contents of the containers are analyzed with regard to their further use. Schenker developed a "Container Algorithm" (CA) for this purpose. This algorithm works both value-oriented and volume-oriented. After it has been activated, the algorithm delivers that container from the dry port to be extracted using the two target criteria.

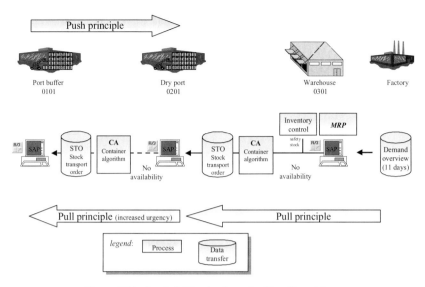

Figure 7.11: Availability check on the Brazilian side

After the executive has confirmed a processing alternative, the containers are transferred from the warehouse into the warehouse of the DC and are removed from the warehouse inventory of the container warehouse. After opening the containers in the warehouse, the inventories are checked again and compared in detail with the notified deliveries of the CC for the first time. Customs duty is paid for the parts, which are then separated and some of then stored as a safety stock. Schenker uses this safety stock to meet unexpected problems, such as damages to parts or other faults, and so avoid a potential production interrupt.

Figure 7.12 illustrates the areas of the DC and describes the goods flow resulting from the information impulses from Mercedes-Benz. The DC receives a JiT schedule from the factory's IT systems in irregular intervals. When the this EDI message arrives, which is imported using a file interface, only 90 minutes remain until the delivery to the factory in the extreme case.

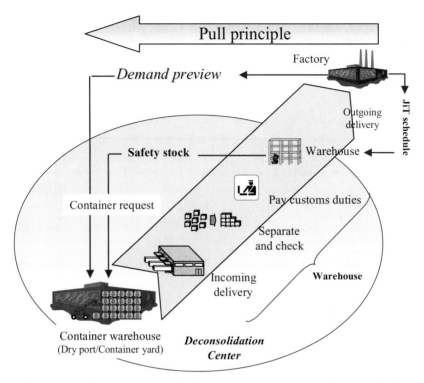

Figure 7.12: Information and goods flows from the container warehouse to the factory

The JiT schedule is converted into deliveries in the R/3-Systems before the goods are delivered by trucks to fixed unloading points near the production conveyor belts. The production or assembly can now start. The confirmation of the arrival, which is not yet passed electronically to the Schenker's R/3 System, causes a final booking and terminates the delivery process.

7.1.6   Overall Summary of the Information-based Logistics Infrastructure

Figure 7.13 then summarizes the information flows that accompany the supply chain. The illustration concentrates on the involved institutes, which primarily concern the application systems and the associated data exchange formats based on the R/3 System.

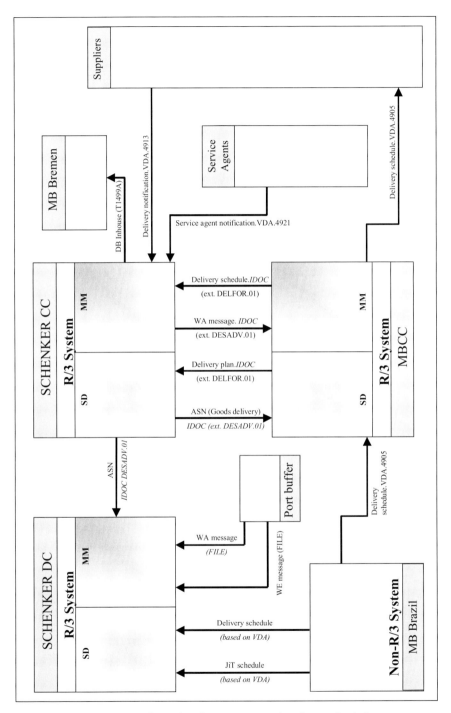

Figure 7.13: Information flows accompanying the supply chain

7.2   Business Information Warehouse Perspective: Inventory Information in the Supply Chain

In addition to the existing services, Schenker Logistics is currently extending its core services. For example, Schenker plans to offer individual customers a summary of all parts available in a supply chain.

Because many stock planners are involved in the monitoring of partial deliveries at both the MBCC and the factory in Brazil, a complementary information service for the described supply chain is explicitly desired from Mercedes-Benz. They continuously check whether the parts planned for a specific production time arrive on schedule and if an alternative planning is possible in case of unexpected events.

Requirements for an inventory information system

The information system must satisfy the following requirements:

- Mercedes-Benz should be given the opportunity to use the Internet to view the current location of all parts in the supply chain and obtain other information about the supply chain.

- The information request should be as simple as possible.

- The instrument should help Schenker to detect problem situations faster and so take account of the associated security aspects better and faster.

- The massive information flow must not affect the operative systems in any way.

- The relevant data of all companies involved with a processing stage should be combined to produce a shared information system. Consequently, the system must permit the integration of various data from heterogeneous systems. The capability to consolidate the information from both Schenker OLTP systems and the data from the shipping companies and the air freight carriers (*Schenker International*) must be provided for the case outlined here (refer to Figure 7.14).

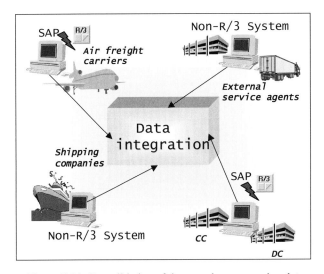

Figure 7.14: Consolidation of the complete processing data

Definition of status points

It is critical for the realization of such a solution that this can be included quickly and without major complications in the frequently changing handling processes and the existing applications. As the following feasibility study and the Schenker SOURCE (Stock Overview Using Relevant Control Events) prototype in this context show, the BW technology and the associated openness compared with external systems provides an instrument that largely meets the specified requirements. The prototype for the inventory information system consists of two major subareas, the BW functionality and the associated data logic for the processing.

In an initial step, four status points were defined for the feasibility study. These items should later by extended to 13. These four status points are

* goods receipt in the CC
* goods issue in the CC
* goods receipt in the DC
* goods issue in the DC.

These states serve as information sources and should form the basis for the SOURCE inventory information system.

Estimation of the data volume

An estimation for the expected data volume and the associated number of data updates serves as basis for the project. The production volume at the factory in Brazil determines the parts contained in the supply chain. On average, approximately 300 vehicles are produced per day. The total average utilization of the supply chain covers 40 production days, namely the parts inventory needed to assemble 12,000 vehicles. Because each vehicle consists of approximately 600 different parts, in the extreme case this represents a maximum data volume of 7,200,000 documents. If a new document is written for each transaction of a part (status change), this would produce a maximum data volume of approximately 94,000,000 document data items for the planned 13 states. On the basis of the estimate of 7,200,000 document data items, which must be continuously maintained in the SOURCE prototype, this produces the need to update 7,500 documents daily. From the BW point of view, this interval can be realized without any difficulty and can even be further reduced.

Definition of InfoCube

The information objects and thus the InfoCubes for the prototypes were defined in a third stage (refer to Section 6.3.3 for information on InfoCubes). As Figure 7.15 shows, these consist of four characteristics and seven key figures. In addition to the materials number and the status, the shipping unit (VSE) number and inward and outward movements form the principal characteristics of the prototype. The VSE number is used to distinguish between two materials with an identical material number within a status.

Because all entries and data changes remain in the facts table, it is necessary to distinguish between inward and outward movements. These use a binary variable to describe the arrival or removal of materials in defined status points; this binary variable can take the value A or R, respectively, for arrival or removal. Together with the time characteristics (calendar year, month and calendar day), they also provide the basis for the analysis of the duration of a material in a status.

In addition to the characteristics, there are also a number of different key figures as shown in the following examples.

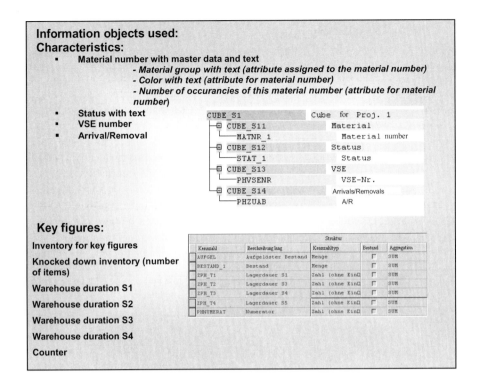

Figure 7.15: Structure of the InfoCube

A counter generated in the facts table is also integrated here. This counter does not result from the transaction data and is used to determine the number of data records. The required material master data was taken from the associated R/3 Systems of the CC and of the DC. It is advantageous that the CC and the DC have identical master data records. Attributes such as color of the material or the material group can be used as detailed information in the queries.

The following example describes the scenario of a data transfer from the R/3 Systems into the BW and illustrates the structure and the handling of the data. The facts table is initially empty.

Example of a data acquisition:

i.   *Read the first data record. This is an arrival with ten units.*

Material	VSE	CALDAY	CALMONTH	STATUS
100	1000	19990101	199901	01

A/R	Inventory
A	10

ii.   *Once the "data upload" has been performed, the resolved inventory items are determined automatically in the facts table. Namely, the total quantity of the material 100 in all VSE. Each VSE contains 10 quantity units for the material. The required information comes from the master data in the Schenker R/3 System.*

Mat.	VSE	CALDAY	CALMONTH	Stat.	A/R	Inventory	Knocked down inv.
100	1000	19990101	199901	01	A	10	100

S1	S2	S3	S4	C
0	0	0	0	1

All characteristics are identical with the loaded data record. In contrast, the key figures are extended with the resolved inventory (lot size), the actual wait times and the previously mentioned counter.

iii.   *A second "data upload" moves the material from status 1 to status 2.*

Material	VSE	CALDAY	CALMONTH	STATUS
100	1000	19990103	199901	01
100	1000	19990103	199901	02

A/R	Inventory
R	-10
A	10

iv.   *This is represented in the facts table by a removal (R) in status 1 and an arrival (A) in status 2.*

Mat.	VSE	CALDAY	CALMONTH	Stat.	A/R	Inventory	Knocked down inv.
100	1000	19990101	199901	01	A	10	100
100	1000	19990103	199901	01	R	-10	-100
100	1000	19990103	199901	02	A	10	100

S1	S2	S3	S4	C
0	0	0	0	1
2	0	0	0	1
0	0	0	0	1

Thus, status 1 is credited from the removed quantity of material 100. Because all data records initially remain in the facts table, this affects three records.

As the example shows, no entries are deleted from the facts table. After the second data upload, an offsetting entry (-10) for the material passed to status 2 (10) is made to status 1. Status 2 is debited accordingly.

The duration of storage of a material in a status needed for the time analyses result from the difference between the removal and arrival date in a status. Because the calendar month and the day must also be updated, it is also possible to analyze whether the transport within the supply chain goes over a month boundary. The prototype currently provides only the capability to perform day analyses, for example to answer the question how many days a material remains at a status point. An extension of this restriction is planned. The counter C in each data line makes it possible to use historic values to calculate an average wait time for the number of read records.

Evaluation and analysis

The counter and the associated time characteristic provides Schenker Logistics and Mercedes-Benz with the opportunity to determine the expected duration of any material in a status. These durations result from historic values. Mercedes-Benz has the advantage for its planning that a material does not need to be present at a status point in order to analyze its average duration. This means that queries do not depend on the availability of a material in a status.

Figure 7.16 shows one of the many opportunities for reports and analysis of the data material. In the example selected here, the "material inventory" key figure depends on different states and times. This makes it possible, for example, to query the inventory in a specific month in a specific status. Figure 7.16 shows 11 of the 13 planned status points.

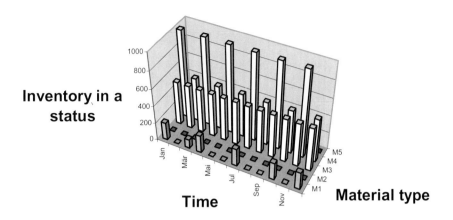

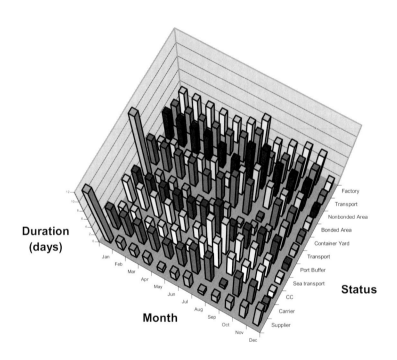

Figure 7.16: Graphical reports

Additional reports and analyses provide Mercedes-Benz with the possibility to use the expected duration for their own planning.

Reports for the end-user

End-users login to the SOURCE system and use the Business Explorer, to get access to a number of different queries. As already described in Section 6.3.3, users can store queries in the Business Explorer and then start these after entering a material number using the channel display. Figure 7.17 shows how the user can request information for specific materials in defined status points.

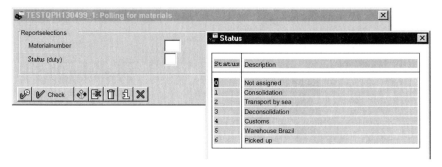

Figure 7.17: Execution of a query

After initiating a query, the end-user obtains, for example, the report shown in Figure 7.18 in Microsoft Excel format.

⅀ SCHENKER
Stinnes Logistics

Materialnumber Status					
Last update	23.05.00 08:59				
Mateial number	100	101	102	110	111
	Airbag, steering wheel	Airbag, doors	Airbag, side	Bumper, front	Bumper, rear
Status	1	1	1	1	1
	Consolidation	Consolidation	Consolidation	Consolidation	Consolidation
stock	3 pcs	5 pcs	4 pcs	6 pcs	10 pcs
knocked down inventory	30 pcs	50 pcs	80 pcs	30 pcs	50 pcs
exp. duration time	0	0	0	0	0
exp. duration time in status 1	0	0	0	0	0

Figure 7.18: Report after the first data transfer (status 1)

The screenshot in Figure 7.18 shows an overview of all available materials in a status. In addition to the inventory, the user gets an independent view of all available parts. Because this is the first data transfer, no duration times have been calculated. The Scheduler for the Administrator Workbench can be used to load new data records periodically or event-controlled.

ℤ SCHENKER
Stinnes Logistics

Materialnumber						
Status						

Last update	23.05.00 09:01					
Mateial number	100	101	102	110	111	
	Airbag, steering wheel	Airbag, doors	Airbag, side	Bumper, front	Bumper, rear	
Status	1	1	1	1	1	
	Consolidation	Consolidation	Consolidation	Consolidation	Consolidation	
stock	8 pcs	6 pcs	10 pcs	0	0	
knocked down inventory	80 pcs	60 pcs	200 pcs	0	0	
exp. duration time	1,3	0,7	1,3	3	2	
exp. duration time in status 1	1,3	0,7	1,3	3	2	

Figure 7.19: Report after the second data transfer (status 1)

In contrast, Figure 7.19 shows a corresponding overview after the second data upload. Because the screenshot in Figure 7.19 shows only status 1, the materials with identification number larger than the value 102 that were transferred into status 2 are no longer visible. An arrival is also indicated for the other materials (100-102).

Assessment

The described prototype shows that the BW can be used to realize an inventory information system. In addition to the individual status summary, it is also possible to analyze the material distribution over various states. In case you encounter a problem one may use a faster transport medium (e.g. airplane). In this case the Information System allows to omit predefined status points.

Rudimentary InfoCubes already provide a number of analysis possibilities for the solution. These include:

- analysis of the complete inventories with regard to status, month, material or material group
- trends for the duration of storage with regard to status, month, material or material group
- calculation of the frequency of transshipment
- identification of missing stock (provided appropriate planned values have been specified)
- queries correct to the day over the complete acquired time period
- analysis of inventory at the VSE level (item level).

If a "Container" characteristic is added (analog to VSE), all queries can also be used for the container.

The described examples for the basis for a number of Schenker-internal benchmarks serve to show the saving potential for both costs and time, and also to make use of these.

The aggregation level of an InfoCube is a critical success factor for the query performance. If they do not consist more than 6,000 data entries, response times are less than five seconds.

The querier receives an acceptable performance when the InfoCubes aggregate is maintained. Because all entries remain in the facts table, its updating is critical. Because the massive data volumes can quickly increase its size, old data packets are removed from the facts table before reconstructing an aggregate. However, this does not mean that these data are deleted from the core system. In principle, it can also be used in other InfoCubes. Although the elimination of obsolete data records is only theoretically possible, it is not considered to be an unsolvable problem.

Because the front-end must currently be installed on the querier's systems, only solutions of external providers can be used. In addition to an Internet capability from both parties, comprehensive graphical analyses characterize these, although at additional cost.

The SOURCE system should provide the customers with other value-added services in addition to a comprehensive, transparent display. For example, supplier assessments and a number of different ABC analyses should be possible over the Internet.

Chapter 8 Goodyear Case Study: Use of SAP Systems for Supply Chain Management

The Goodyear Tire & Rubber Company with 95,000 employees worldwide and a daily production of more than 500,000 tires is one of the largest tire and rubber manufacturers in the world. The company has itself six rubber plantations and maintains 84 factories in 26 countries (refer to Figure 8.1). Goodyear has development centers in the USA, Japan and Luxembourg for research and development.

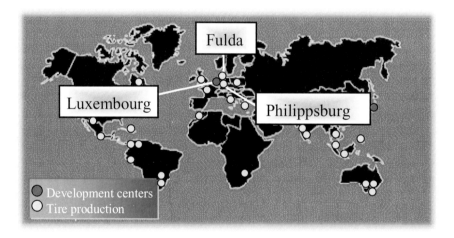

Figure 8.1: Worldwide positioning

The company operates with the brands Goodyear, Fulda, Kelly, Lee, Delta, Voyager, Debica and Sava, and is represented in 42 countries in which the Goodyear products are marketed with over 25,000 sales branches. In Germany more than 60,000 distribution locations are supplied with products of the Goodyear Group.

In addition to the large automobile manufacturers, franchise holders and service companies, the customers include both large and small companies such as service stations. The Goodyear Group is represented in Germany at the two locations, Fulda and Philippsburg. These are also part of our case study. In addition to a central R/3 System physically located Luxembourg, Goodyear uses a distributed IT architecture comprising of APO, LES and BW for the logistics processing (refer to Figure 8.2).

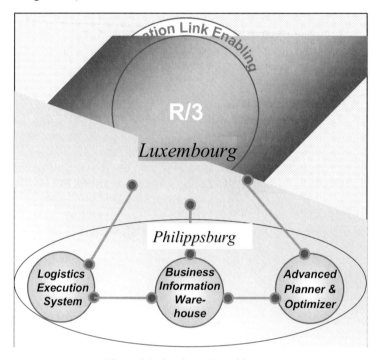

Figure 8.2: Goodyear IT architecture

The SAP LES was developed in cooperation with Goodyear, which was also one of the twelve APO pilot customers. Goodyear also runs a BW, and with the distributed system architecture, uses a wide spectrum from the SAP Business Framework. The following sections describe the use of these systems in conjunction with three case studies.

8.1 APO

8.1.1 Task

Goodyear has various locations worldwide that produce specific types of tires and are managed in independent SAP systems. Because the individual factories do not provide the complete Goodyear product range, a mutual product transfer on the basis of a cooperative planning is needed to supply the markets. Figure 8.3 shows the SAP systems at the various concern locations.

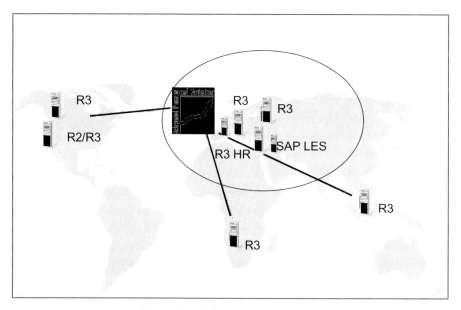

Figure 8.3: Global system landscape

The APO aims at linking different central systems for the planning and control of the goods flow between different locations. The scope is to reduce the costs for inventory, handling and transport by 20-30 percent, to speed up the order processing by 25 percent, and to improve the product availability and delivery by more than 95 percent. In particular, the individual locations should have all needed resources when they are required.

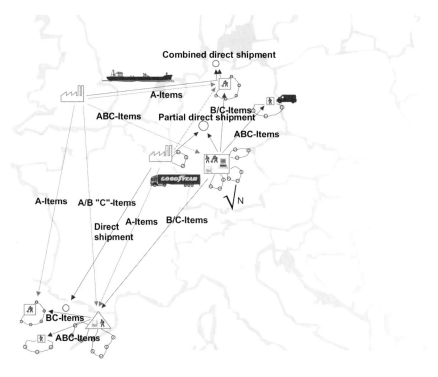

Figure 8.4: Delivery strategies

As Figure 8.4 shows, there are several delivery and distribution alternatives to supply different locations with goods in Goodyear's European Network. The simultaneous planning needed for an optimization of the logistics processing over country boundaries requires a common view of the various orders over all instances of the participating systems. For the transport management, this means the determination of the optimum route, the selection of the transport equipment and the transport type, and the determination of the needed loading sequences from a general perspective. For example, it should be possible to use various factors such as cost and time to develop an optimum from a central view. This implies a dynamic delivery strategy between locations.

8.1.2   Use of APO Components

This section describes the use of APO components for the previously represented task. The foundation for the Demand Planning, Supply Network Planning and Deployment, Production Planning, and Available to Promise were implemented in an initial phase. The SCC displayed in Figure 8.5 shows an overview of all locations integrated within the model and their relationships.

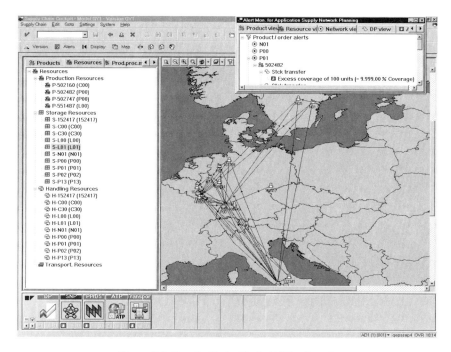

Figure 8.5: Supply Chain Cockpit (Goodyear)

The Application Link Enabling (ALE) technology was used to link three European central systems with the APO. ALE designates an SAP-specific technology for the integration of loosely-coupled, independent applications with physically separated R/3 Systems. This is a medium for the asynchronous coupling of distributed systems without a shared database. The communication within ALE is performed using messages and is established with Remote Function Calls (RFC) between two applications. ALE also has the advantage that separate R/3 Systems with different release levels – as in the case of Goodyear – can communicate with each other.

Each individual central system administers a specific number of distribution centers, plants, delivery warehouses and consignment warehouses that have been modeled using the Supply Chain Engineer.

The model contains a number of restrictions. These are

- production restrictions,
- transport restrictions, and
- warehouse restrictions.

The following section introduces the APO planning modules used.

Demand Planning

The Demand planning module includes and evaluates in the company planning not only order data, but also information of the MM module. The Demand Planning uses seasonal trend models, multilinear regression and exponential smoothing with the existing inventory levels to estimate the future sales. Various factors, such as extreme winter weather, registration figures for cars and trucks, test results from an automobile journal and their effects on the demand for special products are automatically integrated in the forecast of the planning module.

The sales forecast determines the quantity and revenues that Goodyear attempts to achieve using a specific marketing plan and in accordance with the expected market situation. The user can configure the planning horizon and the historical data horizon used as basis, for example weeks, months or years.

Supply Network Planning and Deployment

The forecast is passed to the Supply Network Planning and Deployment module. The forecast of the product demand is solved using heuristics and assigned to the various locations.

As Figure 8.6 shows, the user receives a summary of all available inventory, safety stock, potential supply bottlenecks and distribution information. Supply bottlenecks can be avoided by using production and procurement plans, which are forwarded to the appropriate factories. From a plant viewpoint, a production plan represents the request to produce a certain quantity of one or more products by a certain time. If there is also a requirement for raw materials here, the required supplier, which can also be a Goodyear plant, is informed. The results of this planning run are transferred to the associated R/3 Systems as a scheduling agreement or production plan. These are confirmed in the systems of the suppliers or factories after making the corresponding Detailed Scheduling.

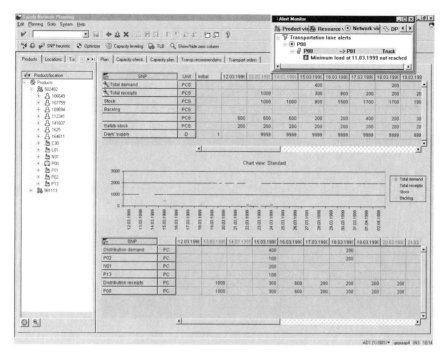

Figure 8.6: Supply Network Planning

The results of the planning are forwarded to the Transport Load Builder, which performs a cross-system transport optimization and calculates how and in which form the deliveries are to be made.

Production Planning and Detailed Scheduling

Previously Goodyear used the production planning module only to inform individual locations and systems which products and quantities are required at a certain time. Functions of the detailed planning, for example the assignments of production orders to machines, are not used.

Available to Promise

Rule-based criteria can be used to determine even at the availability check how the supply to locations and the customer is to be made. The linking of different systems provides Goodyear with the capability to make availability requests not only to country-specific systems, but also at a worldwide cross-system level. An ATP cost rate has been defined as restriction, which, when exceeded, causes the rejection of a delivery. In future, the ATP functionality should be extended to production orders and transport orders.

8.1.3   SAP APO – An Assessment

The implementation started in November 1998 and resulted in June 1999 with the productive start for some selected products. Difficulties were mainly concerned with the complex integration work through to the mutual exchange of information between the APO and the participating R/3 systems. A second implementation phase started at the beginning of 2000 with Release 2.0. It should cover the use of further functionality of the individual modules, the previously neglected Detailed Scheduling and aspects of the Strategic Planning. The latter should make it possible to check the efficiency of the individual locations using simulations and, for example, analyze alternative location decisions and their effects on cost, time and effort.

The first results of using the APO indicate that significant cost savings have been achieved in the complete European network. Extrapolated to the implementation of the complete product range, savings of 25-30% in the inventory area (stock costs) and 20-30% in the area of the transport costs can be expected. The ATP standard scenarios previously implemented for the three European central systems produced a significant reduction of the response times, both for external customers and also for concern-internal queries.

8.2   Support for Logistics Processes Using the SAP Logistics Execution System

8.2.1   Initial Situation and Task

Goodyear has two central installations in Germany that required an instrument to improve the processing in the warehouse and transport areas. The first installation is the "Factory Warehouse" in Fulda (Germany), which is a factory and warehouse with a sortiment of A, B and C-articles. The second installation is a location in Philippsburg (Germany) which produces goods and also serves as distribution center for A, B and C-products made worldwide. For example, tires produced in Asia are distributed from Philippsburg to other locations of the Goodyear Group or to customers.

The management of these two locations was initially realized within a central R/3

System in Luxembourg. However, this resulted in the following problems:

- The high data arrival rate meant that the realization within the R/3 architecture was possible only to a limited extent, because, depending on the loading, the response times of the central system as result of the data storage on a database could be very long.

- Updates to the central system, for example because of release changes, affect the availability of the systems at both locations.

- The functionality for the support of decentral warehouse and transport processes provided with the R/3 System did not suffice for an efficient support of these business processes at both locations.

With this background, a SAP pilot project with the implementation of a decentral system physically located in Philippsburg was started in February 1997. The aim was to retain the central order management in Luxembourg and with new functionality provide long-term support for the warehouse and transport processes at the two locations. As Figure 8.7 shows, the R/3 applications were connected using ALE.

The LES manages the Fulda and Philippsburg locations since September 1997. The logistics processing in Philippsburg is currently supported using the Mobile Operating on Business (MOB) forklift guiding system from Witron. The LES supports the warehouse management, transport and distribution functions within the R/3 System in Luxembourg. As seen from the two locations, this decentral system physically located in Philippsburg has two major advantages. It is available 24 hours per day, seven days per week, and is independent of the central system.

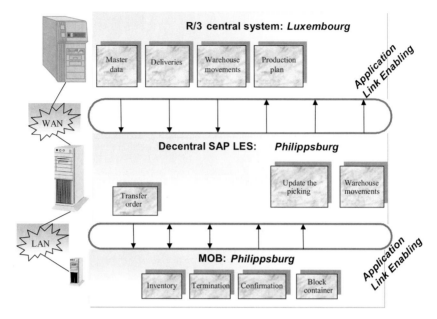

Figure 8.7: Goodyear system architecture

The following section discusses the transportation management and warehouse management, two central processes in logistics, and their support provided by the LES.

8.2.2   Transport Management with LES

Orders are managed using the central system in Luxembourg. In general, incoming orders are differentiated into large customers and day-to-day business.

The large customer business handles, for example, orders from large automobile concerns. The orders are bound to outline agreements and are characterized by a large volume. JiT deliveries are characteristic for this type of business. A forecast schedule is added automatically by EDI to the central system in Luxembourg and made concrete up to four hours before the car manufacturer's production with a JiT schedule (refer to Figure 8.8).

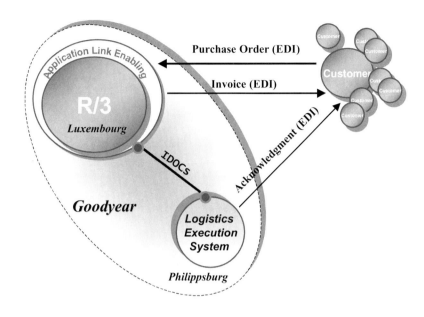

Figure 8.8: Information exchange for large customer business

Depending on time restrictions and distances, the resulting delivery can be made either from one of the two locations or a consignment warehouse. Acknowledgements, for example delivery notifications, which are currently still handled by the central system, should in future be replaced by the LES.

For the day-to-day business, small companies usually order the required products at the telephone call center in Cologne. These are in most of the cases smaller order quantities. The information is entered there without further delay into the central system. This is necessary because Goodyear provides a 24-hour service from order entry to delivery for this type of business. The current volume for Germany is more than 7,500 orders.

Independent of the type of the received order, deliveries are produced from the orders. A preselection from which location (Fulda or Philippsburg) the delivery to the customer is to be made has already been decided in the central system. Because the physical services, for example the picking, are managed by the LES in Fulda and Philippsburg, a copy of the delivery is passed there. The transmission is performed almost in real-time using the previously discussed ALE technology.

The description of the delivery in the system contains the goods recipient, the order article and the number of units, the total weight, the goods issue date, and all information needed for the further processing (refer to Figure 8.9).

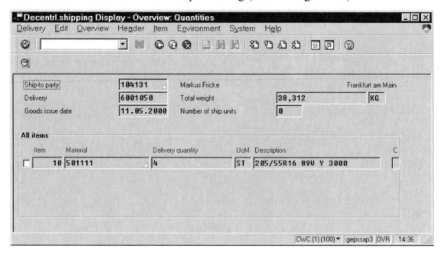

Figure 8.9: Description of the delivery in the decentral system

Whereas for orders from large customers the large volume and thus full loading of the transport medium results in a direct delivery from the plant, the supply of smaller customers usually takes place as a multi-stage transport chain. The individual orders are distributed by various service agents to smaller delivery sites, from where they are then delivered to the customers (refer to Figure 8.10). The transport chain thus consists of a main run (to the delivery warehouse) which is then followed by a local distribution (to the customer). The partner companies are responsible for the coordination of the local distribution. To ensure that the 24-hour service is maintained, the service agents involved with the processing are assigned to fixed gates or ramps and time windows. This matches the loading and delivery times so that the delivery warehouse, and thus the customers, can be supplied on time.

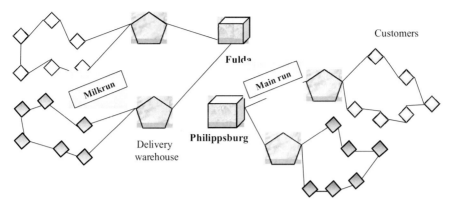

Figure 8.10: Delivery in day-to-day business

The 24-hour service offered by Goodyear requires a comprehensive shipment disposition for deliveries with trucks. For this purpose, shipment orders, called shipments in the SAP terminology, that contain all the information needed for the shipment processing are created. As Figure 8.11 shows, these contain information about

- the shipment route
- the service agent
- the loading gate
- the loading times, and
- the start and end of the shipment.

The system automatically assigns the shipment route, service agent and loading gate. However, these are only suggestions oriented on rule-based criteria and which the user can change at any time. Information about the loading times is needed to control the subsequent picking of the deliveries.

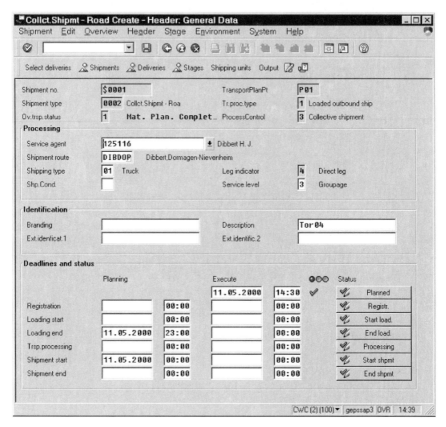

Figure 8.11: Assemble stop-off shipments

The deliveries are automatically assigned to other vehicles if the delivery volume exceeds the capacity of a vehicle. A so-called "greedy algorithm" and the use of other algorithms perform a simple sorting that continues to assign different deliveries to a transport medium until its capacity is reached. The algorithm starts a new run as soon as the capacity of the first vehicle is exhausted. This assignment ends when all deliveries have been assigned to the available transport medium. Figure 8.12 illustrates this with an example in which the deliveries are assigned to three different means of shipment for the same service agent.

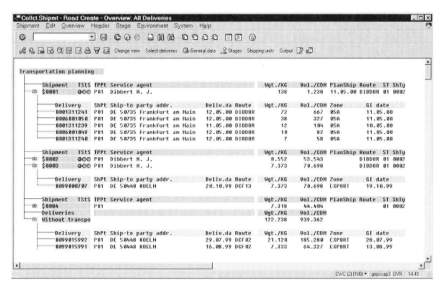

Figure 8.12: Assignment of deliveries to three different means of transport

8.2.3   Order Picking with LES

The order picking is initiated at the end of assembling the stop-off shipment. The aim here is to reduce the activities within the warehouse and thus minimize the associated work.

The LES supports a two-step picking. This picking is not restricted to a single delivery, i.e., not oriented to customers and individual deliveries, and provides the capability to pick products from different deliveries for an order in a single step. This means that a consignor sees just the tires to be picked and not the associated order. Thus, for example, he can select all identical products of a shipment in a single step. The LES provides a suggestion that makes an assignment in "picking waves". These specify which tires are to be picked when and by whom. As Figure 8.13 shows, the various items of a delivery for a specific customer may well be assigned in different waves. The form of the picking shown is based on a individual and project-specific modification that was developed by Goodyear and SAP together; the SAP standard provides just a delivery or order-related picking.

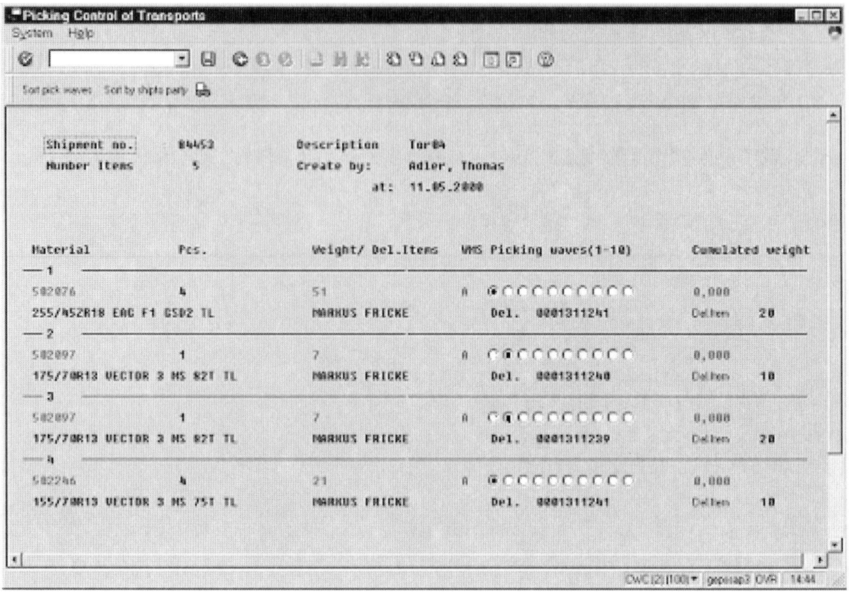

Figure 8.13: Different picking waves

The assignment and organization of deliveries and their constituent items are made using rule-based criteria. The rule type contains, for example, information about the warehouse personnel and the available equipment, such as forklift trucks. The size, volume and warehouse position are used to determine which equipment is needed for the picking. For example, a distinction can be made here between picking in bulk storage and picking in a shelf warehouse. Depending on the warehouse location, different equipment is needed to perform the processing. Because the personnel also has different qualifications, each product cannot be "picked" by any arbitrary employee. Figure 8.13 shows that the items of a shipment can be picked in different waves.

After the individual products have been assigned during different picking waves, these data are transferred using ALE from the LES into a *forklift guiding system.* This is currently the previously mentioned system from Witron, which to support the picking, forwards suggestions directly to radio terminals of the forklift truck drivers. The system shows the priorities of the shipments and which warehouse locations contain the associated tires. It also determines the shortest path through the warehouse. The electronic confirmation of the removed tire is made directly in the forklift truck. The status after each confirmation is transferred to the LES and provides an overview at the earliest possible time. The labeling of the tires is performed in the same sequence in which the shipment orders were picked, and initiated directly by the forklift truck driver's activities.

A control station functionality that was specially developed by Goodyear provides the capability to represent the progress of the order picking and any problems that occur there.

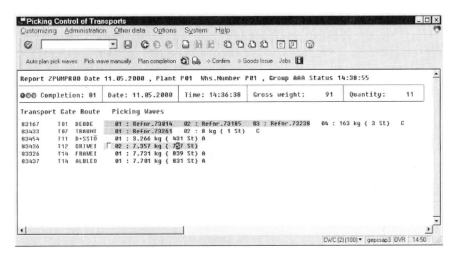

Figure 8.14: Order picking status

As Figure 8.14 shows, it is possible to determine, which processes have the waiting, processing or fault status. Items shown with a light gray background (yellow in the R/3 System) and an A-marking (in Figure 8.14, for example numbers T83326 and T83437) describe shipments yet to be processed and indicate to the warehouse personnel which processes are time-critical. In contrast, a C-marking indicates that faults occurred during the processing (number T83167, item 04, and number T83433, item 02). Items shown with a dark gray background (green in the R/3 System) have the processing status (for example number T83436 or number T83167, items 01-03). The visualization at the control station makes it possible to interact with the warehouse personnel and so correct any problems at an early stage.

A delivery-related goods issue entry is made after the loading has been completed. Each individual delivery is passed as an IDOC to the central system and initiates there an EDI transfer to the customer or the distribution center. These are acknowledgements and invoices. In future, the confirmations will be able to be made directly from the decentral system.

8.2.4   SAP LES – An Assessment

From Goodyear's point of view, the implementation of a self-contained LES produced the following results:

- The effort for the system administration was reduced to 20 minutes per month and satisfied the expectations made for the LES.

- The personnel costs for the maintenance and updating of the system could be significantly reduced in comparison to the old system.

- No hardware or software errors occurred during the more than two years of operations.

- The average response times, which, depending on the loading, were extremely long due to the data administration by the central system, could be reduced by the factor three through the transition to the LES.

- The personnel and warehouse planning has been simplified.

- The commissioning of the LES has resulted in an improvement in productivity. Specifically, 30 percent more deliveries could be handled with the same number of personnel.

- The times for the assembly and picking of the deliveries are less than two hours and have been reduced by approximately 60 percent.

- The handling effort for the loading and unloading of a transport medium was affected only to a small extent and lies under ten minutes.

- The fault incidence has been reduced and thus the quality of the in-house processes improved.

- The forklift guiding system from Witron, in productive use since mid-1998, has realized further saving potential of ten percent for personnel and throughput.

8.3 Use of the SAP Business Information Warehouse at Goodyear

To administer and prepare the comprehensive information of the two systems previously discussed, the system architecture in Philippsburg has been extended by SAP BW. This runs on a Windows NT computer with a 2-processor machine and two gigabytes of main memory. The BW has the following main uses:

- remove mass data from the operative systems, such as the created deliveries and shipments, and

- provide data to perform analyses and reports.

The massive data volume has already lowered the performance of the operative systems after two years by approximately 50 percent. Because the mass data must not have any affect on the operative daily processing, a transfer has been made to the BW. This elimination of data from the operative systems has restored the original performance. In addition, the administered information can now be viewed at any time in the BW and used in a subsequent company planning.

Together with the performance improvement, the BW is primarily used to perform comprehensive analyses and reports. As Figure 8.15 shows, Goodyear has defined several different InfoCubes for this purpose. These cover the business areas ranging from procurement, include warehouse management, quality management, the material requirements planning and sales, through to cost control.

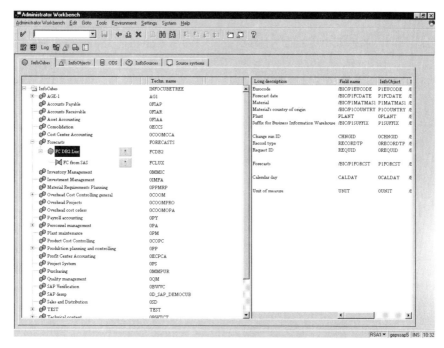

Figure 8.15: Different InfoCubes

Together with the LES, it can be used, for example, to investigate in detail the vehicle capacity loading, the workload, the assignment of the products in the warehouse or the order picking.

The BW is also a useful addition to the APO by permitting data to be transferred from the APO to the BW. These are order and delivery data, and forecasts from the Demand Planning module. If all data records were held permanently in the APO, this would result in an unsatisfactory system performance and thus no longer provide real-time information.

Abbreviations

ABAP	Advanced Business Application Programming
AGate	Application Gateway
ALE	Application Link Enabling
ANSI	American National Standards Institute
APO	Advanced Planner and Optimizer
ASN	Advanced Shipping Notification
ASRS	Automatic Storage and Retrieval Systems
ATP	Available to Promise
BAPIs	Business Application Programming Interfaces
BC	Business Connector
DIN	Deutsches Institut für Normung e.V (German Industrial Standards Institute)
BW	Business Information Warehouse
CAD	Computer Aided Design
CAS	Computer Aided Selling
CC	Consolidation Center
CIC	Customer Interaction Center
CLMS	Car Location Message Systems
COM/DCOM	Common Object Model/Distributed Common Object Model
CORBA	Common Object Request Broker Architecture
COSA	Cooperative Simulated Annealing
CPFR	Collaborative Planning, Forecasting and Replenishment
CRM	Customer Relationship Management
DB	Database
DC	Deconsolidation Center
DTD	Document Type Definition
EAN	European Article Number
EC	Electronic Commerce
Ecash	Electronic Cash
EDI	Electronic Data Interchange

EDIFACT	Electronic Data Interchange for Administration, Commerce and Transport
ETA	Estimated Time of Arrival
ftp	File Transfer Protocol
GA	Genetic algorithms
GPS	Global Positioning System
HTML	Hypertext Markup Language
HTTP	Hypertext Transfer Protocol
IACs	Internet Application Components
ID	Identification
IDOCs	Intermediate Documents
IP	Internet Protocol
ISDN	Integrated Services Digital Network
IT	Information Technology
ITS	Internet Transaction Server
IP	Information processing
JiT	Just-in-Time
LAN	Local Area Network
LE	Logistics Execution
LES	Logistics Execution System
LIS	Logistics Information System
MBCC	Mercedes-Benz – Consolidation Center
MB – Brazil	Mercedes-Benz – Brazil
MM	Materials Management
MOB	Mobile Operating on Business
MRP	Material Required Planning
ODS	Operational Data Store
OEM	Original Equipment Manufacturer
OLAP	Online Analytical Processing
OLTP	Online Transaction Processing
P-Cards	Purchasing Cards
PDM	Product Data Management
PIN	Personal Identification Number
PM	Plant Maintenance
PP	Production Planning
PS	Project System
QM	Quality Management
RFC	Remote Function Call
SA	Simulated Annealing
SAP	Systems, Applications and Products
SAPGUI	SAP Graphical User Interface
SCC	Supply Chain Cockpit
SCE	Supply Chain Engineer
SCE	Sales Configuration Engine

SCOREx	Supply Chain Optimization and Real-Time Extended Executions
SCS	Supply Chain Solutions
SD	Sales and Distribution
SET	Secure Electronic Transaction
SFA	Sales Force Automation
SM	Service Management
smtp	Simple Mail Transfer Protocol
SOURCE	Stock Overview Using Relevant Control Events
SPE	Sales Pricing Engine
SSL	Secure Socket Layer
STO	Stock Transfer Order
SWIFT	Society for Worldwide Interbank Financial Telecommunication
TMS	Transportation Management System
TCP/IP Protocol	Transmission Control Protocol/Internet Protocol
TRADACOMS	Trading Data Communications Standards
VANs	Value Added Networks
WGate	Web-gateway
WM	Warehouse Management
WMS	Warehouse Management System
WWW	World Wide Web
XML	Extensible Markup Language

Bibliography

Appelrath, H.-J./Ritter, J. (2000): SAP R/3 Implementation – Methods and Tools. Springer, Berlin, etc.

Boyes, W./Melvin, M. (1994): Microeconomics. 2nd Edition. Houghton Mifflin Company, Boston, Toronto etc.

Buxmann, P./Gebauer, J. (1999): Evaluating the Use of Information Technology in Inter-Organizational Relationships. In: Proceedings of the 32nd Hawaii International Conference on System Sciences.

Chapman, D. Brent/Zwicky, E. D. (1995): Building internet firewalls. O'Reilly, Cambridge, etc.

Dearing, B. (1990): The Strategic Benefits of EDI. In: The Journal of Business, Jan./Feb., pp. 4-6.

Domschke, W./Drexl, A. (1996): Logistik – Standorte. 4th Edition. Oldenbourg, Munich and Vienna.

Gehring, H./Menschner, K./Meyer, M. (1990): A computer-based heuristic for packing pooled shipment containers. In: European Journal of Operational Research, Vol. 44, pp. 277-288.

Gravelle, H./Rees, R. (1992): Microeconomics. 2nd Edition. Longman Group. New York.

Hadley, G./Whitin, T. M. (1963): Analysis of Inventory Systems. Englewood Cliffs.

Hulihahn, J. B. (1985): International Supply Chain Management. In: International Journal of Physical Distribution and Materials Management, Vol. 15, No. 1, pp. 51-66.

Ihde, G. B. (1991): Transport, Verkehr, Logistik – gesamtwirtschaftliche Aspekte und einzelwirtschaftliche Handhabung. 2th Edition. Vahlen, Munich.

Isermann, H. (1998a): Grundlagen eines systemorientierten Logistikmanagements. In: Isermann, H. (editor) Logistik – Die Gestaltung von Logistiksystemen. 2th Edition. Verlag Moderne Industrie, Landsberg/Lech, pp. 1-42.

Isermann, H. (1998b): Stauraumplanung. In: Isermann, H. (editor) Logistik – Die Gestaltung von Logistiksystemen. 2th Edition. Verlag Moderne Industrie, Landsberg/Lech, pp. 245-286.

Jarke, M./Lenzerini, M./Vassiliou, Y./Vassiliadis, P. (2000): Fundamentals of data warehouses. Springer, Berlin, Heidelberg, New York etc.

Keller, G./Teufel, T. (1998): SAP R/3 Process Oriented Implementation: Iterative Process Prototyping. Addison-Wesley, Bonn, etc.

Knolmayer, G./Mertens, P./Zeier, A. (2000): Supply Chain Management Based on SAP Systems – Order Processing in Manufacturing Companies. Springer, Berlin, etc.

Krieger, W. (1995): Informationsmanagement in der Logistik – Grundlagen – Anwendungen – Wirtschaftlichkeit. Gabler, Wiesbaden.

Lackes, R. (1996): Kanban. In: Kern, W. etc. (editor): Handwörterbuch der Produktionswirtschaft, 2th Edition. Schäffer-Poeschel, Stuttgart, pp. 839-852.

Mertens, P. (1995): Supply Chain Management (SCM). In: Wirtschaftsinformatik, Volume 37, No. 2, pp. 177-179.

Pérez, M./Hildebrand, A./Matzke, B./Zencke, P. (1999): SAP R/3 System on the Internet. Addison-Wesley, Bonn, etc.

Pfohl, H.-C. (1990): Logistiksysteme – Betriebswirtschaftliche Grundlagen. 4th Edition. Springer, Berlin, etc.

Plattner, H. (1999): Customer Relationship Management. In: Scheer, A.-W./Nüttgens, M. (editor): Electronic Business Engineering. Physica-Verlag, Heidelberg, pp. 1-12.

Rebel, T./Darge, O./König, W. (1997): Approaches of Digital Signature Legislation. Working Paper No. 1997-33. Institut für Wirtschaftinformatik. Frankfurt University.

Rebstock, M./Hildebrand, K. (1999): SAP R/3 Management. The Coriolis Group, Bonn.

Schulte, C. (1995): Logistik – Wege zur Optimierung des Material- und Informationsflusses. 2th Edition. Vahlen, Munich.

Schütte, R. (1997): Supply Chain Management. In: Mertens, P. etc. (editor): Lexikon der Wirtschaftsinformatik. 3th Edition. Springer, Berlin, etc.

Stolpmann, M. (1997): Elektronisches Geld im Internet: Grundlagen, Konzepte, Perspektiven. O'Reilly, Cambridge, etc.

Weitzel, T./Buxmann, P. (2000): A communication architecture for the digital economy – 21st century EDI. In: Proceedings of the 33rd Hawaii International Conference on System Sciences (HICSS-33).

Westarp, F. v./Weitzel, T./Buxmann, P./König, W. (1999): The Status Quo and the Future of EDI. In: Proceedings of European Conference on Information Systems (ECIS).

Wolf, D. (1997): Transportkette. In: Bloech, J., Ihde, G. B. (editor): Vahlens großes Logistiklexikon. Vahlen, Munich, pp. 1089-1093.

Zwass, V. (1996): Electronic Commerce – Structure and Issues. In: International Journal of Electronic Commerce, Vol. 1, No. 1, pp. 3-23.

Index

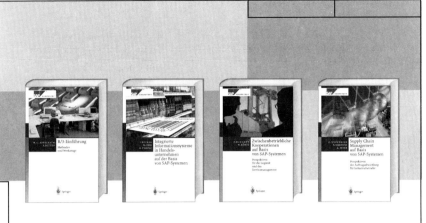

H. Österle, E. Fleisch, R. Alt

Business Networking

Shaping Enterprise Relationships on the Internet with contributions by numerous e,ts

2000. XV, 376 pp. 109 figs., 39 tabs. Hardcover
* DM 98,- ISBN 3-540-66612-5

"The strength of Business Networking is that it brings together historically separated concepts, such as electronic commerce, supply chain management and customer relationship management. The insights of this book are most valuable for our strategies, such as mySAP.com."
Hasso Plattner, Co-Chairman and CEO, SAP AG, Germany

"This book sends an important and timely message; i.e. the nature of business transactions has changed dramatically, and as a consequence business models of competition... must also change. The insights contained in the book are varied and are based on a strong grounding of leading edge practice."
Morris A. Cohen, Professor, The Wharton School, University of Pennsilvania, USA

A.-W. Scheer

ARIS - Business Process Frameworks

3rd ed. 1999. XVII, 186 pp. 94 figs. Hardcover
* DM 79,- 3-540-65834-3

The importance of the link between business process organization and strategic management is stressed. Bridging the gap between the different approaches in business theory and information technology, the ARIS concept provides a full-circle approach - from the organizational design of business processes to IT implementation.

ARIS - Business Process Modeling

3rd ed. 2000. XIX, 218 pp. 179 figs. Hardcover
* DM 89,- ISBN 3-540-65835-1

This book describes in detail how ARIS methods model and realize business processes by means of UML (Unified Modeling Language), leading to an information model that is the keystone for a systematic and intelligent method of developing application systems.
Multiple real-world examples - including knowledge management, implementation of workflow systems and standard software solutions (SAP R/3 in particular) - address the deployment of ARIS methods.